Functional Programming

by John Paul Mueller

A Wiley Brand

Functional Programming For Dummies®

Published by: **John Wiley & Sons, Inc.,** 111 River Street, Hoboken, NJ 07030-5774, www.wiley.com

Copyright © 2019 by John Wiley & Sons, Inc., Hoboken, New Jersey

Published simultaneously in Canada

For general information on our other products and services, please contact our Customer Care Department within the U.S. at 877-762-2974, outside the U.S. at 317-572-3993, or fax 317-572-4002. For technical support, please visit https://hub.wiley.com/community/support/dummies.

Wiley publishes in a variety of print and electronic formats and by print-on-demand. Some material included with standard print versions of this book may not be included in e-books or in print-on-demand. If this book refers to media such as a CD or DVD that is not included in the version you purchased, you may download this material at http://booksupport.wiley.com. For more information about Wiley products, visit www.wiley.com.

Library of Congress Control Number: 2018965285

ISBN: 978-1-119-52750-3

ISBN 978-1-119-52751-0 (ebk); ISBN ePDF 978-1-119-52749-7 (ebk)

Manufactured in the United States of America

C10007288_010419

Table of Contents

Introduction

The *functional programming paradigm* is a framework that expresses a particular set of assumptions, relies on particular ways of thinking through problems, and uses particular methodologies to solve those problems. Some people view this paradigm as being akin to performing mental gymnastics. Other people see functional programming as the most logical and easiest method for coding any particular problem ever invented. Where you appear in this rather broad range of perspectives depends partly on your programming background, partly on the manner in which you think through problems, and partly on the problem you're trying to solve.

Functional Programming For Dummies doesn't try to tell you that the functional programming paradigm will solve every problem, but it does help you understand that functional programming can solve a great many problems with fewer errors, less code, and a reduction in development time. Most important, it helps you understand the difference in the thought process that using the functional programming paradigm involves. Of course, the key is knowing when functional programming is the best option, and that's what you take away from this book. Not only do you see how to perform functional programming with both pure (Haskell) and impure (Python) languages, but you also gain insights into when functional programming is the best solution.

About This Book

Functional Programming For Dummies begins by describing what a paradigm is and how the functional programming paradigm differs. Many developers today don't really understand that different paradigms can truly change the manner in which you view a problem domain, thereby making some problem domains considerably easier to deal with. As part of considering the functional programming paradigm, you install two languages: Haskell (a pure functional language) and Python (an impure functional language). Of course, part of this process is to see how pure and impure languages differ and determine the advantages and disadvantages of each.

Part of working in the functional programming environment is to understand and use lambda calculus, which is part of the basis on which functional programming it built. Imagine that you're in a room with some of the luminaries of computer science and they're trying to decide how best to solve problems in computer science at a time when the term *computer science* doesn't even exist. For that matter, no one has even defined what it means to compute. Even though functional programming might seem new to many people, it's based on real science created by the best minds the world has ever seen to address particularly difficult problems. This science uses lambda calculus as a basis, so an explanation of this particularly difficult topic is essential.

After you understand the basis of the functional programming paradigm and have installed tools that you can use to see it work, it's time to create some example code. This book starts with some relatively simple examples that you might find in other books that use other programming paradigms so that you compare them and see how functional programming actually differs. You then move on to other sorts of programming problems that begin to emphasize the benefits of functional programming in a stronger way. To make absorbing the concepts of functional programming even easier, this book uses the following conventions:

>> Text that you're meant to type just as it appears in the book is **bold**. The exception is when you're working through a step list: Because each step is bold, the text to type is not bold.

>> Because functional programming will likely seem strange to many of you, I've made a special effort to define terms, even some of those that you might already know, because they may have a different meaning in the functional realm. You see the terms in italics, followed by their definition.

>> When you see words in *italics* as part of a typing sequence, you need to replace that value with something that works for you. For example, if you see "Type **Your Name** and press Enter," you need to replace *Your Name* with your actual name.

>> Web addresses and programming code appear in mono font. If you're reading a digital version of this book on a device connected to the Internet, note that you can click the web address to visit that website, like this: www.dummies.com.

>> When you need to type command sequences, you see them separated by a special arrow, like this: File ⇨ New File. In this case, you go to the File menu first and then select the New File entry on that menu. The result is that you see a new file created.

Foolish Assumptions

You might find it difficult to believe that I've assumed anything about you — after all, I haven't even met you yet! Although most assumptions are indeed foolish, I made these assumptions to provide a starting point for the book.

You need to be familiar with the platform that you want to use because the book doesn't provide any guidance in this regard. To give you maximum information about the functional programming paradigm, this book doesn't discuss any platform-specific issues. You need to know how to install applications, use applications, and generally work with your chosen platform before you begin working with this book. Chapter 2 does show how to install Python, and Chapter 3 shows how to install Haskell. Part 2 of the book gives you the essential introduction to functional programming, and you really need to read it thoroughly to obtain the maximum benefit from this book.

This book also assumes that you can find things on the Internet. Sprinkled throughout are numerous references to online material that will enhance your learning experience. However, these added sources are useful only if you actually find and use them.

Icons Used in This Book

As you read this book, you see icons in the margins that indicate material of interest (or not, as the case may be). This section briefly describes each icon in this book.

TIP

Tips are nice because they help you save time or perform some task without a lot of extra work. The tips in this book are time-saving techniques or pointers to resources that you should try in order to get the maximum benefit from Python, Haskell, or the functional programming paradigm.

WARNING

I don't want to sound like an angry parent or some kind of maniac, but you should avoid doing anything marked with a Warning icon. Otherwise, you could find that your program serves only to confuse users, who will then refuse to work with it.

TECHNICAL STUFF

Whenever you see this icon, think advanced tip or technique. You might find these tidbits of useful information just too boring for words, or they could contain the solution that you need to get a program running. Skip these bits of information whenever you like.

REMEMBER

If you don't get anything else out of a particular chapter or section, remember the material marked by this icon. This text usually contains an essential process or a bit of information that you must know to write Python, Haskell, or functional programming applications successfully.

Beyond the Book

This book isn't the end of your functional programming experience — it's really just the beginning. I provide online content to make this book more flexible and better able to meet your needs. That way, as I receive email from you, I can do things like address questions and tell you how updates to Python, its associated packages, Haskell, it's associated libraries, or changes to functional programming techniques that affect book content. In fact, you gain access to all these cool additions:

>> **Cheat sheet:** You remember using crib notes in school to make a better mark on a test, don't you? You do? Well, a cheat sheet is sort of like that. It provides you with some special notes about tasks that you can do with Python or Haskell that not every other developer knows. In addition, you find some quick notes about functional programming paradigm differences. You can find the cheat sheet for this book by going to www.dummies.com and searching this book's title. Scroll down the page until you find a link to the Cheat Sheet.

>> **Updates:** Sometimes changes happen. For example, I might not have seen an upcoming change when I looked into my crystal ball during the writing of this book. In the past, that simply meant the book would become outdated and less useful, but you can now find updates to the book by searching this book's title at www.dummies.com.

In addition to these updates, check out the blog posts with answers to reader questions and demonstrations of useful book-related techniques at http://blog.johnmuellerbooks.com/.

>> **Companion files:** Hey! Who really wants to type all the code in the book? Most readers would prefer to spend their time actually working through coding examples, rather than typing. Fortunately for you, the source code is available for download, so all you need to do is read the book to learn functional programming techniques. Each of the book examples even tells you precisely which example project to use. You can find these files at www.dummies.com. Click More about This Book and, on the page that appears, scroll down the page to the set of tabs. Click the Downloads tab to find the downloadable example files.

Where to Go from Here

It's time to start your functional programming paradigm adventure! If you're a complete functional programming novice, you should start with Chapter 1 and progress through the book at a pace that allows you to absorb as much of the material as possible.

If you're a novice who's in an absolute rush to get going with functional programming techniques as quickly as possible, you can skip to Chapter 2, followed by Chapter 3, with the understanding that you may find some topics a bit confusing later. You must install both Python and Haskell to have any hope of getting something useful out of this book, so unless you have both languages installed, skipping these two chapters will likely mean considerable problems later.

Readers who have some exposure to functional programming and already have both Python and Haskell installed can skip to Part 2 of the book. Even with some functional programming experience, Chapter 5 is a must-read chapter because it provides the basis for all other discussions in the book. The best idea is to at least skim all of Part 2.

If you're absolutely certain that you understand both functional programming paradigm basics and how lambda calculus fits into the picture, you can skip to Part 3 with the understanding that you may not see the relevance of some examples. The examples build on each other so that you gain a full appreciation of what makes the functional programming paradigm different, so try not to skip any of the examples, even if they seem somewhat simplistic.

1
Getting Started with Functional Programming

Discover the functional programming paradigm.

Understand how functional programming differs.

Obtain and install Python.

Obtain and install Haskell.

Chapter **1**

Introducing Functional Programming

This book isn't about a specific programming language; it's about a programming paradigm. A *paradigm* is a framework that expresses a particular set of assumptions, relies on particular ways of thinking through problems, and uses particular methodologies to solve those problems. Consequently, this programming book is different because it doesn't tell you which language to use; instead, it focuses on the problems you need to solve. The first part of this chapter discusses how the functional programming paradigm accomplishes this task, and the second part points out how functional programming differs from other paradigms you may have used.

The math orientation of functional programming means that you might not create an application using it; you might instead solve straightforward math problems or devise *what if* scenarios to test. Because functional programming is unique in its approach to solving problems, you might wonder how it actually accomplishes its goals. The third section of this chapter provides a brief overview of how you use the functional programming paradigm to perform various kinds of tasks (including traditional development), and the fourth section tells how some languages follow a pure path to this goal and others follow an impure path. That's not to say that those following the pure path are any more perfect than those following the impure path; they're simply different.

Finally, this chapter also discusses a few online resources that you see mentioned in other areas of the book. The functional programming paradigm is popular for solving certain kinds of problems. These resources help you discover the specifics of how people are using functional programming and why they feel that it's such an important method of working through problems. More important, you'll discover that many of the people who rely on the functional programming paradigm aren't actually developers. So, if you aren't a developer, you may find that you're already in good company by choosing this paradigm to meet your needs.

Defining Functional Programming

Functional programming has somewhat different goals and approaches than other paradigms use. Goals define what the functional programming paradigm is trying to do in forging the approaches used by languages that support it. However, the goals don't specify a particular implementation; doing that is within the purview of the individual languages.

REMEMBER

The main difference between the functional programming paradigm and other paradigms is that functional programs use math functions rather than statements to express ideas. This difference means that rather than write a precise set of steps to solve a problem, you use math functions, and you don't worry about how the language performs the task. In some respects, this makes languages that support the functional programming paradigm similar to applications such as MATLAB. Of course, with MATLAB, you get a user interface, which reduces the learning curve. However, you pay for the convenience of the user interface with a loss of power and flexibility, which functional languages do offer. Using this approach to defining a problem relies on the *declarative programming* style, which you see used with other paradigms and languages, such as Structured Query Language (SQL) for database management.

In contrast to other paradigms, the functional programming paradigm doesn't maintain state. The use of *state* enables you to track values between function calls. Other paradigms use state to produce variant results based on environment, such as determining the number of existing objects and doing something different when the number of objects is zero. As a result, calling a functional program function always produces the same result given a particular set of inputs, thereby making functional programs more predictable than those that support state.

Because functional programs don't maintain state, the data they work with is also *immutable*, which means that you can't change it. To change a variable's value, you must create a new variable. Again, this makes functional programs more predictable than other approaches and could make functional programs easier to run on multiple processors. The following sections provide additional information on how the functional programming paradigm differs.

Understanding its goals

Imperative programming, the kind of programming that most developers have done until now, is akin to an assembly line, where data moves through a series of steps in a specific order to produce a particular result. The process is fixed and rigid, and the person implementing the process must build a new assembly line every time an application requires a new result. Object-oriented programming (OOP) simply modularizes and hides the steps, but the underlying paradigm is the same. Even with modularization, OOP often doesn't allow rearrangement of the object code in unanticipated ways because of the underlying interdependencies of the code.

REMEMBER

Functional programming gets rid of the interdependencies by replacing procedures with pure functions, which requires the use of immutable state. Consequently, the assembly line no longer exists; an application can manipulate data using the same methodologies used in pure math. The seeming restriction of immutable state provides the means to allow anyone who understands the math of a situation to also create an application to perform the math.

Using pure functions creates a flexible environment in which code order depends on the underlying math. That math models a real-world environment, and as our understanding of that environment changes and evolves, the math model and functional code can change with it — without the usual problems of brittleness that cause imperative code to fail. Modifying functional code is faster and less error prone because the person implementing the change must understand only the math and doesn't need to know how the underlying code works. In addition, learning how to create functional code can be faster as long as the person understands the math model and its relationship to the real world.

Functional programming also embraces a number of unique coding approaches, such as the capability to pass a function to another function as input. This capability enables you to change application behavior in a predictable manner that isn't possible using other programming paradigms. As the book progresses, you encounter other such benefits of using functional programming.

Using the pure approach

Programming languages that use the pure approach to the functional programming paradigm rely on lambda calculus principles, for the most part. In addition, a pure-approach language allows the use of functional programming techniques only, so that the result is always a functional program. The pure-approach language used in this book is Haskell because it provides the purest implementation, according to articles such as the one found on Quora at `https://www.quora.com/ What-are-the-most-popular-and-powerful-functional-programming- languages`. Haskell is also a relatively popular language, according to the TIOBE index (`https://www.tiobe.com/tiobe-index/`). Other pure-approach languages include Lisp, Racket, Erlang, and OCaml.

WARNING

As with many elements of programming, opinions run strongly regarding whether a particular programming language qualifies for pure status. For example, many people would consider JavaScript a pure language, even though it's untyped. Others feel that domain-specific declarative languages such as SQL and Lex/Yacc qualify for pure status even though they aren't general programming languages. Simply having functional programming elements doesn't qualify a language as adhering to the pure approach.

Using the impure approach

Many developers have come to see the benefits of functional programming. However, they also don't want to give up the benefits of their existing language, so they use a language that mixes functional features with one of the other programming paradigms (as described in the "Considering Other Programming Paradigms" section that follows). For example, you can find functional programming features in languages such as C++, C#, and Java. When working with an impure language, you need to exercise care because your code won't work in a purely functional manner, and the features that you might think will work in one way actually work in another. For example, you can't pass a function to another function in some languages.

TIP

At least one language, Python, is designed from the outset to support multiple programming paradigms (see `https://blog.newrelic.com/2015/04/01/ python-programming-styles/` for details). In fact, some online courses make a point of teaching this particular aspect of Python as a special benefit (see `https:// www.coursehero.com/file/p1hkiub/Python-supports-multiple-programming- paradigms-including-object-oriented/`). The use of multiple programming paradigms makes Python quite flexible but also leads to complaints and apologists (see `http://archive.oreilly.com/pub/post/pythons_weak_functional_progra. html` as an example). The reasons that this book relies on Python to demonstrate the impure approach to functional programming is that it's both popular and flexible, plus it's easy to learn.

Considering Other Programming Paradigms

You might think that only a few programming paradigms exist besides the functional programming paradigm explored in this book, but the world of development is literally packed with them. That's because no two people truly think completely alike. Each paradigm represents a different approach to the puzzle of conveying a solution to problems by using a particular methodology while making assumptions about things like developer expertise and execution environment. In fact, you can find entire sites that discuss the issue, such as the one at `http://cs.lmu.edu/~ray/notes/paradigms/`. Oddly enough, some languages (such as Python) mix and match compatible paradigms to create an entirely new way to perform tasks based on what has happened in the past.

REMEMBER

The following sections discuss just four of these other paradigms. These paradigms are neither better nor worse than any other paradigm, but they represent common schools of thought. Many languages in the world today use just these four paradigms, so your chances of encountering them are quite high.

Imperative

Imperative programming takes a step-by-step approach to performing a task. The developer provides commands that describe precisely how to perform the task from beginning to end. During the process of executing the commands, the code also modifies application state, which includes the application data. The code runs from beginning to end. An imperative application closely mimics the computer hardware, which executes machine code. *Machine code* is the lowest set of instructions that you can create and is mimicked in early languages, such as assembler.

Procedural

Procedural programming implements imperative programming, but adds functionality such as code blocks and procedures for breaking up the code. The compiler or interpreter still ends up producing machine code that runs step by step, but the use of procedures makes it easier for a developer to follow the code and understand how it works. Many procedural languages provide a disassembly mode in which you can see the correspondence between the higher-level language and the underlying assembler. Examples of languages that implement the procedural paradigm are C and Pascal.

TECHNICAL STUFF

Early languages, such as Basic, used the imperative model because developers creating the languages worked closely with the computer hardware. However, Basic users often faced a problem called *spaghetti code,* which made large applications appear to be one monolithic piece. Unless you were the application's developer, following the application's logic was often hard. Consequently, languages that follow the procedural paradigm are a step up from languages that follow the imperative paradigm alone.

Object-oriented

The procedural paradigm does make reading code easier. However, the relationship between the code and the underlying hardware still makes it hard to relate what the code is doing to the real world. The object-oriented paradigm uses the concept of objects to hide the code, but more important, to make modeling the real world easier. A developer creates code objects that mimic the real-world objects they emulate. These objects include properties, methods, and events to allow the object to behave in a particular manner. Examples of languages that implement the object-oriented paradigm are C++ and Java.

REMEMBER

Languages that implement the object-oriented paradigms also implement both the procedural and imperative paradigms. The fact that objects hide the use of these other paradigms doesn't mean that a developer hasn't written code to create the object using these older paradigms. Consequently, the object-oriented paradigm still relies on code that modifies application state, but could also allow for modifying variable data.

Declarative

Functional programming actually implements the declarative programming paradigm, but the two paradigms are separate. Other paradigms, such as logic programming, implemented by the Prolog language, also support the declarative programming paradigm. The short view of declarative programming is that it does the following:

›› Describes what the code should do, rather than how to do it

›› Defines functions that are referentially transparent (without side effects)

›› Provides a clear correspondence to mathematical logic

Using Functional Programming to Perform Tasks

It's essential to remember that functional programming is a paradigm, which means that it doesn't have an implementation. The basis of functional programming is lambda calculus (https://brilliant.org/wiki/lambda-calculus/), which is actually a math abstraction. Consequently, when you want to perform tasks by using the functional programming paradigm, you're really looking for a programming language that implements functional programming in a manner that meets your needs. (The next section, "Discovering Languages that Support Functional Programming," describes the available languages in more detail.) In fact, you may even be performing functional programming tasks in your current language without realizing it. Every time you create and use a lambda function, you're likely using functional programming techniques (in an impure way, at least).

In addition to using lambda functions, languages that implement the functional programming paradigm have some other features in common. Here is a quick overview of these features:

» **First-class and higher-order functions:** First-class and higher-order functions both allow you to provide a function as an input, as you would when using a higher-order function in calculus.

» **Pure functions:** A pure function has no side effects. When working with a pure function, you can

- Remove the function if no other functions rely on its output

- Obtain the same results every time you call the function with a given set of inputs

- Reverse the order of calls to different functions without any change to application functionality

- Process the function calls in parallel without any consequence

- Evaluate the function calls in any order, assuming that the entire language doesn't allow side effects

» **Recursion:** Functional language implementations rely on recursion to implement looping. In general, recursion works differently in functional languages because no change in application state occurs.

» **Referential transparency:** The value of a variable (a bit of a misnomer because you can't change the value) never changes in a functional language implementation because functional languages lack an assignment operator.

REMEMBER

You often find a number of other considerations for performing tasks in functional programming language implementations, but these issues aren't consistent across languages. For example, some languages use strict (eager) evaluation, while other languages use non-strict (lazy) evaluation. Under strict evaluation, the language fully checks the function before evaluating it. Even when a term within the function isn't used, a failing term will cause the function as a whole to fail. However, under non-strict evaluation, the function fails only if the failing term is used to create an output. The Miranda, Clean, and Haskell languages all implement non-strict evaluation.

Various functional language implementations also use different type systems, so the manner in which the underlying computer detects the type of a value changes from language to language. In addition, each language supports its own set of data structures. These kinds of issues aren't well defined as part of the functional programming paradigm, yet they're important to creating an application, so you must rely on the language you use to define them for you. Assuming a particular implementation in any given language is a bad idea because it isn't well defined as part of the paradigm.

Discovering Languages That Support Functional Programming

To actually use the functional programming paradigm, you need a language that implements it. As with every other paradigm discussed in this chapter, languages often fall short of implementing every idea that the paradigm provides, or they implement these ideas in unusual ways. Consequently, knowing the paradigm's rules and seeing how the language you select implements them helps you to understand the pros and cons of a particular language better. Also, understanding the paradigm makes comparing one language to another easier. The functional programming paradigm supports two kinds of language implementation, pure and impure, as described in the following sections.

Considering the pure languages

A pure functional programming language is one that implements only the functional programming paradigm. This might seem a bit limited, but when you read through the requirements in the "Using Functional Programming to Perform Tasks" section, earlier in the chapter, you discover that functional programming is mutually exclusive to programming paradigms that have anything to do with the imperative paradigm (which applies to most languages available today).

Trying to discover which language best implements the functional programming paradigm is nearly impossible because everyone has an opinion on the topic. You can find a list of 21 functional programming language implementations with their pros and cons at `https://www.slant.co/topics/485/~best-languages-for-learning-functional-programming`.

Considering the impure languages

Python is likely the epitome of the impure language because it supports so many coding styles. That said, the flexibility that Python provides is one reason that people like using it so much: You can code in whatever style you need at the moment. The definition of an impure language is one that doesn't follow the rules for the functional programming paradigm fully (or at least not fully enough to call it pure). For example, allowing any modification of application state would instantly disqualify a language from consideration.

REMEMBER

One of the more common and less understood reasons for disqualifying a language as being a pure implementation of the functional programming paradigm is the lack of pure-function support. A pure function defines a specific relationship between inputs and outputs that has no side effects. Every call to a pure function with specific inputs always garners precisely the same output, making pure functions extremely reliable. However, some applications actually rely on side effects to work properly, which makes the pure approach somewhat rigid in some cases. Chapters 4 and 5 provide specifics on the question of pure functions. You can also discover more in the article at `http://www.onlamp.com/2007/07/12/introduction-to-haskell-pure-functions.html`.

Finding Functional Programming Online

Functional programming has become extremely popular because it solves so many problems. As covered in this chapter, it also comes with a few limitations, such as an inability to use mutable data; however, for most people, the pros outweigh the cons in situations that allow you to define a problem using pure math. (The lack of mutable data support also has pros, as you discover later, such as an ability to perform multiprocessing with greater ease.) With all this said, it's great to have resources when discovering a programming paradigm. This book is your first resource, but a single book can't discuss everything.

TIP

Online sites, such as Kevin Sookochef (`https://sookocheff.com/post/fp/a-functional-learning-plan/`) and Wildly Inaccurate (`https://wildlyinaccurate.com/functional-programming-resources/`), offer a great many helpful resources. Hacker News (`https://news.ycombinator.com/item?id=16670572`) and Quora (`https://www.quora.com/What-are-good-resources-for-teaching-children-functional-programming`) can also be great resources. The referenced Quora site is especially important because it provides information that's useful in getting children started with functional programming. One essential aspect of using online sites is to ensure that they're timely. The resource shouldn't be more than two years old; otherwise, you'll be getting old news.

Sometimes you can find useful videos online. Of course, you can find a plethora of videos of varying quality on YouTube (`https://www.youtube.com/results?search_query=Functional+Programming`), but don't discount sites, such as tinymce (`https://go.tinymce.com/blog/talks-love-functional-programming/`). Because functional programming is a paradigm and most of these videos focus on a specific language, you need to choose the videos you watch with care or you'll get a skewed view of what the paradigm can provide (as contrasted with the language).

WARNING

One resource that you can count on being biased are tutorials. For example, the tutorial at `https://www.hackerearth.com/practice/python/functional-programming/functional-programming-1/tutorial/` is all about Python, which, as noted in previous sections of this chapter, is an impure implementation. Likewise, even solid tutorial makers, such as Tutorials Point (`https://www.tutorialspoint.com/functional_programming/functional_programming_introduction.htm`), have a hard time with this topic because you can't demonstrate a principle without a language. A tutorial can't teach you about a paradigm — at least, not easily, and not much beyond an abstraction. Consequently, when viewing a tutorial, even a tutorial that purports to provide an unbiased view of functional programming (such as the one at `https://codeburst.io/a-beginner-friendly-intro-to-functional-programming-4f69aa109569`), count on some level of bias because the examples will likely appear using a subset of the available languages.

IN THIS CHAPTER

» Obtaining and using Python

» Downloading and installing the datasets and example code

» Running an application

» Writing Python code

Chapter **2**

Getting and Using Python

As mentioned in Chapter 1, Python is a flexible language that supports multiple coding styles, including an implementation of the functional programming paradigm. However, Python's implementation is impure because it does support the other coding styles. Consequently, you choose between flexibility and the features that functional programming can provide when you choose Python. Many developers choose flexibility (and therefore Python), but there is no right or wrong choice — just the choice that works best for you. This chapter helps you set up, configure, and become familiar with Python so that you can use it in the book chapters that follow.

WARNING

This book uses Anaconda 5.1, which supports Python 3.6.4. If you use a different distribution, some of the procedural steps in the book will likely fail to work as expected, the screenshots will likely differ, and some of the example code may not run. To get the maximum benefit from this book, you need to use Anaconda 5.1, configured as described in the remainder of this chapter. The example application and other chapter features help you test your installation to ensure that it works as needed, so following the chapter from beginning to end is the best idea for a good programming experience.

Working with Python in This Book

You could download and install Python 3.6.4 to work with the examples in this book. Doing so would still allow you to gain an understanding of how functional programming works in the Python environment. However, using the pure Python installation will also increase the amount of work you must perform to have a good coding experience and even potentially reduce the amount you learn because your focus will be on making the environment work, rather than seeing how Python implements the functional programming paradigm. Consequently, this book relies on the Jupyter Notebook Integrated Development Environment (IDE) (or user interface or editor, as you might prefer) of the Anaconda tool collection to perform tasks for the reasons described in the following sections.

Creating better code

A good IDE contains a certain amount of intelligence. For example, the IDE can suggest alternatives when you type the incorrect keyword, or it can tell you that a certain line of code simply won't work as written. The more intelligence that an IDE contains, the less hard you have to work to write better code. Writing better code is essential because no one wants to spend hours looking for errors, called *bugs.*

IDEs vary greatly in the level and kind of intelligence they provide, which is why so many IDEs exist. You may find the level of help obtained from one IDE to be insufficient to your needs, but another IDE hovers over you like a mother hen. Every developer has different needs and, therefore, different IDE requirements. The point is to obtain an IDE that helps you write clean, efficient code quickly and easily.

Debugging functionality

Finding bugs (errors) in your code involves a process called *debugging.* Even the most expert developer in the world spends time debugging. Writing perfect code on the first pass is nearly impossible. When you do, it's cause for celebration because it won't happen often. Consequently, the debugging capabilities of your IDE are critical. Unfortunately, the debugging capabilities of the native Python tools are almost nonexistent. If you spend any time at all debugging, you quickly find the native tools annoying because of what they don't tell you about your code.

The best IDEs double as training tools. Given enough features, an IDE can help you explore code written by true experts. Tracing through applications is a time-honored method of learning new skills and honing the skills you already possess. A seemingly small advance in knowledge can often become a huge savings in time later. When looking for an IDE, don't just look at debugging features as a means to remove errors — see them also as a means to learn new things about Python.

Defining why notebooks are useful

Most IDEs look like fancy text editors, and that's precisely what they are. Yes, you get all sorts of intelligent features, hints, tips, code coloring, and so on, but at the end of the day, they're all text editors. Nothing is wrong with text editors, and this chapter isn't telling you anything of the sort. However, given that Python developers often focus on scientific applications that require something better than pure text presentation, using notebooks instead can be helpful.

REMEMBER

A *notebook* differs from a text editor in that it focuses on a technique advanced by Stanford computer scientist Donald Knuth called literate programming. You use *literate programming* to create a kind of presentation of code, notes, math equations, and graphics. In short, you wind up with a scientist's notebook full of everything needed to understand the code completely. You commonly see literate programming techniques used in high-priced packages such as Mathematica and MATLAB. Notebook development excels at

>> Demonstration

>> Collaboration

>> Research

>> Teaching objectives

>> Presentation

This book uses the Anaconda tool collection because it provides you with a great Python coding experience, but also because it helps you discover the enormous potential of literate programming techniques. If you spend a lot of time performing scientific tasks, Anaconda and products like it are essential. In addition, Anaconda is free, so you get the benefits of the literate programming style without the cost of other packages.

Obtaining Your Copy of Anaconda

As mentioned in the previous section, Anaconda doesn't come with your Python installation. With this in mind, the following sections help you obtain and install Anaconda on the three major platforms supported by this book.

Obtaining Analytics Anaconda

The basic Anaconda package comes as a free download that you obtain at `https://www.anaconda.com/download/`. Simply click the symbol for your operating system, such as the window icon for Windows, and then click Download in the platform's section of the page to obtain access to the free product. (Depending on the Anaconda server load, the download can require a while to complete, so you may want to get a cup of coffee while waiting.) Anaconda supports the following platforms:

>> Windows 32-bit and 64-bit (the installer might offer you only the 64-bit or 32-bit version, depending on which version of Windows it detects)

>> Linux 32-bit and 64-bit

>> Mac OS X 64-bit (both graphical and command-line installer)

TIP

You can obtain Anaconda with older versions of Python. If you want to use an older version of Python, click the How to Get Python 3.5 or Other Python Versions link near the middle of the page. You should use an older version of Python only when you have a pressing need to do so, however.

The free product is all you need for this book. However, when you look on the site, you see that many other add-on products are available. These products can help you create robust applications. For example, when you add Accelerate to the mix, you obtain the capability to perform multicore and GPU-enabled operations. The use of these add-on products is outside the scope of this book, but the Anaconda site gives you details on using them.

Installing Anaconda on Linux

You have to use the command line to install Anaconda on Linux; you're given no graphical installation option. Before you can perform the installation, you must download a copy of the Linux software from the Continuum Analytics site. You can find the required download information in the "Obtaining Analytics Anaconda" section, earlier in this chapter. The following procedure should work fine on any Linux system, whether you use the 32-bit or 64-bit version of Anaconda:

1. **Open a copy of Terminal.**

    The Terminal window appears.

2. **Change directories to the downloaded copy of Anaconda on your system.**

    The name of this file varies, but normally it appears as Anaconda3-5.1.0-Linux-x86.sh for 32-bit systems and Anaconda3-5.1.0-Linux-x86_64.sh for 64-bit systems. The version number is embedded as part of the filename. In this case, the filename refers to version 5.1.0, which is the version used for this book. If you use some other version, you may experience problems with the source code and need to make adjustments when working with it.

3. **Type** bash Anaconda3-5.1.0-Linux-x86.sh **(for the 32-bit version) or** bash Anaconda3-5.1.0-Linux-x86_64.sh **(for the 64-bit version) and press Enter.**

An installation wizard starts that asks you to accept the licensing terms for using Anaconda.

4. **Read the licensing agreement and accept the terms using the method required for your version of Linux.**

The wizard asks you to provide an installation location for Anaconda. The book assumes that you use the default location of ~/anaconda. If you choose some other location, you may have to modify some procedures later in the book to work with your setup.

5. **Provide an installation location (if necessary) and press Enter (or click Next).**

The application extraction process begins. After the extraction is complete, you see a completion message.

6. **Add the installation path to your** PATH **statement using the method required for your version of Linux.**

You're ready to begin using Anaconda.

Installing Anaconda on MacOS

The Mac OS X installation comes in only one form: 64-bit. Before you can perform the install, you must download a copy of the Mac software from the Continuum Analytics site. You can find the required download information in the "Obtaining Analytics Anaconda" section, earlier in this chapter.

The installation files come in two forms. The first depends on a graphical installer; the second relies on the command line. The command-line version works much like the Linux version described in the preceding section of this chapter, "Installing Anaconda on Linux.". The following steps help you install Anaconda 64-bit on a Mac system using the graphical installer:

1. **Locate the downloaded copy of Anaconda on your system.**

The name of this file varies, but normally it appears as Anaconda3-5.1.0- MacOSX-x86_64.pkg. The version number is embedded as part of the file- name. In this case, the filename refers to version 5.1.0, which is the version used for this book. If you use some other version, you may experience prob- lems with the source code and need to make adjustments when working with it.

2. **Double-click the installation file.**

An introduction dialog box appears.

3. **Click Continue.**

The wizard asks whether you want to review the Read Me materials. You can read these materials later. For now, you can safely skip the information.

4. **Click Continue.**

The wizard displays a licensing agreement. Be sure to read through the licensing agreement so that you know the terms of usage.

5. **Click I Agree if you agree to the licensing agreement.**

You see a Standard Install dialog box where you can choose to perform a standard installation, change the installation location, or customize your setup. The standard installation is the one you should use for this book. Making changes could cause some steps within the book to fail unless you know how to modify the instructions to suit your setup.

6. **Click Install.**

The installation begins. A progress bar tells you how the installation process is progressing. When the installation is complete, you see a completion dialog box.

7. **Click Continue.**

You're ready to begin using Anaconda.

Installing Anaconda on Windows

Anaconda comes with a graphical installation application for Windows, so getting a good installation means using a wizard, as you would for any other installation. Of course, you need a copy of the installation file before you begin, and you can find the required download information in the "Obtaining Analytics Anaconda" section, earlier in this chapter. The following procedure (which can require a while to complete) should work fine on any Windows system, whether you use the 32-bit or 64-bit version of Anaconda:

1. **Locate the downloaded copy of Anaconda on your system.**

The name of this file varies, but normally it appears as Anaconda3-5.1.0-Windows-x86.exe for 32-bit systems and Anaconda3-5.1.0-Windows-x86_64.exe for 64-bit systems. The version number is embedded as part of the filename. In this case, the filename refers to version 5.1.0, which is the version used for this book. If you use some other version, you may experience problems with the source code and need to make adjustments when working with it.

2. **Double-click the installation file.**

(You may see an Open File – Security Warning dialog box that asks whether you want to run this file. Click Run if you see this dialog box pop up.) You see an Anaconda3 5.1.0 Setup dialog box.

3. **Click Next.**

The wizard displays a licensing agreement. Be sure to read through the licensing agreement so that you know the terms of usage.

4. **Click I Agree if you agree to the licensing agreement.**

You're asked what sort of installation type to perform (personal or for everyone). In most cases, you want to install the product just for yourself. The exception is if you have multiple people using your system and they all need access to Anaconda.

5. **Choose one of the installation types and then click Next.**

The wizard asks where to install Anaconda on disk, as shown in Figure 2-1. The book assumes that you use the default location. If you choose some other location, you may have to modify some procedures later in the book to work with your setup.

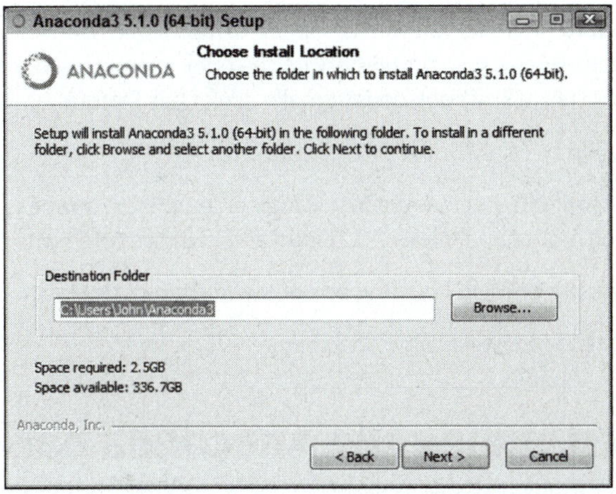

FIGURE 2-1:
Specify an installation location.

6. **Choose an installation location (if necessary) and then click Next.**

You see the Advanced Installation Options, shown in Figure 2-2. These options are selected by default, and no good reason exists to change them in most cases. You might need to change them if Anaconda won't provide your default Python 3.6.4 setup. However, the book assumes that you've set up Anaconda using the default options.

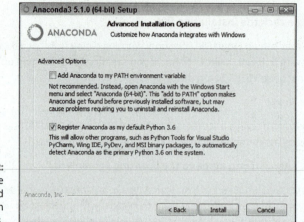

FIGURE 2-2:
Configure the
advanced
installation
options.

7. **Change the advanced installation options (if necessary) and then click Install.**

You see an Installing dialog box with a progress bar. The installation process can take a few minutes, so get yourself a cup of coffee and read the comics for a while. When the installation process is over, you see a Next button enabled.

8. **Click Next.**

The wizard presents you with an option to install Microsoft VSCode. Installing this feature can cause problems with the book examples, so the best idea is not to install it. The book doesn't make use of this feature.

9. **Click Skip.**

The wizard tells you that the installation is complete. You see options for learning more about Anaconda Cloud and getting started with Anaconda.

10. **Choose the desired learning options and then click Finish.**

You're ready to begin using Anaconda.

Understanding the Anaconda package

The Anaconda package contains a number of applications, only one of which you use with this book. Here is a quick rundown on the tools you receive:

>> **Anaconda Navigator:** Displays a listing of Anaconda tools and utilities (installed or not). You can use this utility to install, configure, and launch the various tools and utilities. In addition, Anaconda Navigator provides options to configure the overall Anaconda environment, select a project, obtain help, and interact with the Anaconda community. The "Getting Help with the Python Language" section, at the end of the chapter, tells you more about this tool.

>> **Anaconda Prompt:** Opens a window into which you can type various commands to perform tasks such as starting a tool or utility from the command line, performing installations of sub-features using `pip`, and doing other command line-related tasks.

>> **Jupyter Notebook:** Starts the IDE used for this book. The upcoming "Using Jupyter Notebook" section of the chapter gets you started using the IDE.

>> **Reset Spyder Settings:** Changes the Spyder IDE settings to their original state. Use this option to correct Spyder settings when Spyder becomes unusable or otherwise fails to work as needed.

>> **Spyder:** Starts a traditional IDE that allows you to type source code into an editor window and test it in various ways.

Downloading the Datasets and Example Code

This book is about using Python to perform functional programming tasks. Of course, you can spend all your time creating the example code from scratch, debugging it, and only then discovering how it relates to learning about the wonders of Python, or you can take the easy way and download the prewritten code from the Dummies site as described in the book's Introduction so that you can get right to work.

TIP

To use the downloadable source, you must install Jupyter Notebook. The "Obtaining Your Copy of Anaconda" section, earlier in this chapter, describes how to install Jupyter Notebook as part of Anaconda. You can also download Jupyter Notebook separately from `http://jupyter.org/`. Most of the code in this book will also work with Google Colaboratory, also called Colab (`https://colab.research.google.com/notebooks/welcome.ipynb`), but there is no guarantee all of the examples will work because Colab may not support all the required features and packages. Colab can be handy if you want to work through the examples on your tablet or other Android device. *Python For Data Science For Dummies*, 2nd Edition, by John Paul Mueller and Luca Massaron (Wiley) contains an entire chapter about using Colab with Python and can give you additional help.

The following sections show how to work with Jupyter Notebook, one of the tools found in the Anaconda package. These sections emphasize the capability to manage application code, including importing the downloadable source and exporting your amazing applications to show friends.

Using Jupyter Notebook

To make working with the code in this book easier, you use Jupyter Notebook. This IDE lets you easily create Python notebook files that can contain any number of examples, each of which can run individually. The program runs in your browser, so which platform you use for development doesn't matter; as long as it has a browser, you should be okay.

Starting Jupyter Notebook

Most platforms provide an icon to access Jupyter Notebook. Just click this icon to access Jupyter Notebook. For example, on a Windows system, you choose Start ⇨ All Programs ⇨ Anaconda 3 ⇨ Jupyter Notebook. Figure 2-3 shows how the interface looks when viewed in a Firefox browser. The precise appearance on your system depends on the browser you use and the kind of platform you have installed.

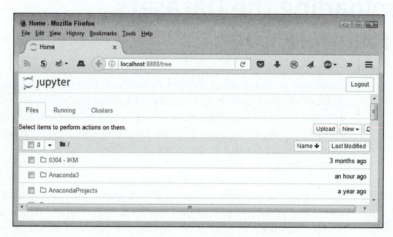

FIGURE 2-3:
Jupyter Notebook provides an easy method to create machine learning examples.

Stopping the Jupyter Notebook server

No matter how you start Jupyter Notebook (or just Notebook, as it appears in the remainder of the book), the system generally opens a command prompt or terminal window to host Jupyter Notebook. This window contains a server that makes the application work. After you close the browser window when a session is complete, select the server window and press Ctrl+C or Ctrl+Break to stop the server.

Defining the code repository

The code you create and use in this book will reside in a repository on your hard drive. Think of a *repository* as a kind of filing cabinet where you put your code. Notebook opens a drawer, takes out the folder, and shows the code to you. You can modify it, run individual examples within the folder, add new examples, and simply interact with your code in a natural manner. The following sections get you started with Notebook so that you can see how this whole repository concept works.

Defining the book's folder

It pays to organize your files so that you can access them more easily later. This book keeps its files in the FPD (Functional Programming For Dummies) folder. Use these steps within Notebook to create a new folder:

1. **Choose New ⇨ Folder.**

Notebook creates a new folder named Untitled Folder. The file appears in alphanumeric order, so you may not initially see it. You must scroll down to the correct location.

2. **Select the box next to the Untitled Folder entry.**

3. **Click Rename at the top of the page.**

You see a Rename Directory dialog box like the one shown in Figure 2-4.

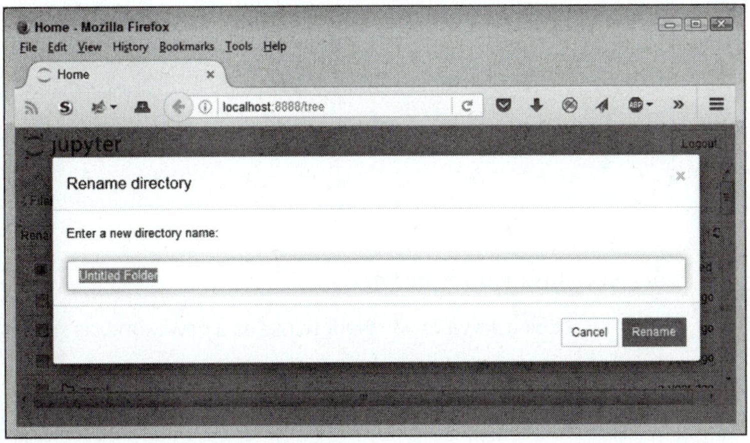

FIGURE 2-4:
Rename the folder so that you remember the kinds of entries it contains.

4. **Type** FPD **and click Rename.**

Notebook changes the name of the folder for you.

5. **Click the new FPD entry in the list.**

Notebook changes the location to the FPD folder in which you perform tasks related to the exercises in this book.

Creating a new notebook

Every new notebook is like a file folder. You can place individual examples within the file folder, just as you would sheets of paper into a physical file folder. Each example appears in a cell. You can put other sorts of things in the file folder, too, but you see how these things work as the book progresses. Use these steps to create a new notebook:

1. Click New ⇨ Python 3.

A new tab opens in the browser with the new notebook, as shown in Figure 2-5. Notice that the notebook contains a cell and that Notebook has highlighted the cell so that you can begin typing code in it. The title of the notebook is Untitled right now. That's not a particularly helpful title, so you need to change it.

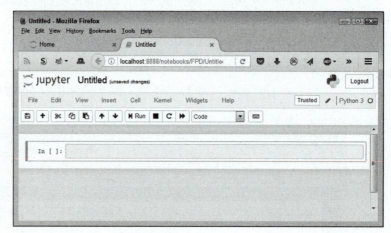

FIGURE 2-5: A notebook contains cells that you use to hold code.

2. Click Untitled on the page.

Notebook asks what you want to use as a new name, as shown in Figure 2-6.

3. Type FPD_02_Sample **and press Enter.**

The new name tells you that this is a file for *Functional Programming For Dummies,* Chapter 2, Sample.ipynb. Using this naming convention lets you easily differentiate these files from other files in your repository.

Of course, the Sample notebook doesn't contain anything just yet. Place the cursor in the cell, type **print('Python is really cool!')**, and then click the Run button. You see the output shown in Figure 2-7. The output is part of the same cell as the code (the code resides in a square box and the output resides outside that square box, but both are within the cell). However, Notebook visually separates the output from the code so that you can tell them apart. Notebook creates a new cell for you.

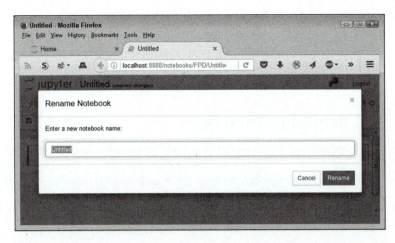

FIGURE 2-6:
Provide a new name for your notebook.

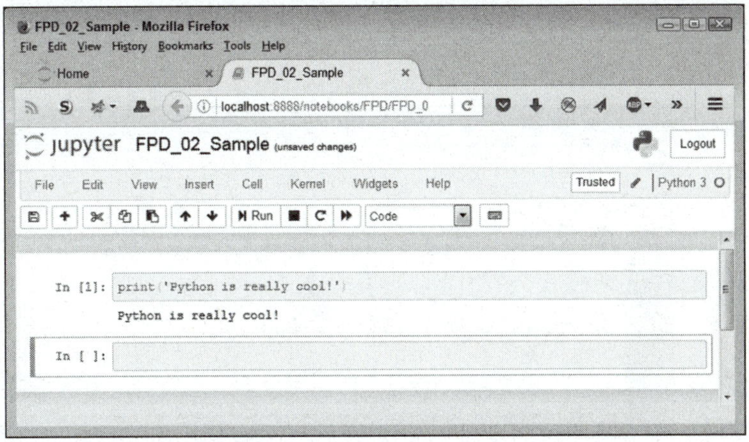

FIGURE 2-7:
Notebook uses cells to store your code.

When you finish working with a notebook, shutting it down is important. To close a notebook, choose File⇨Close and Halt. You return to Notebook's Home page, where you can see that the notebook you just created is added to the list.

Exporting a notebook

Creating notebooks and keeping them all to yourself isn't much fun. At some point, you want to share them with other people. To perform this task, you must export your notebook from the repository to a file. You can then send the file to someone else, who will import it into a different repository.

The previous section shows how to create a notebook named `FPD_02_Sample.ipynb` in Notebook. You can open this notebook by clicking its entry in the repository list. The file reopens so that you can see your code again. To export this code, choose File⇨Download As⇨Notebook (.ipynb). What you see next depends on your browser, but you generally see some sort of dialog box for saving the notebook as a file. Use the same method for saving the Notebook file as you use for any other file you save by using your browser. Remember to choose File⇨Close and Halt when you finish so that the application shuts down.

Removing a notebook

Sometimes notebooks get outdated or you simply don't need to work with them any longer. Rather than allow your repository to get clogged with files that you don't need, you can remove these unwanted notebooks from the list. Use these steps to remove the file:

1. Select the box next to the FPD_02_Sample.ipynb entry.

2. Click the trash can icon (Delete) at the top of the page.

You see a Delete notebook warning message like the one shown in Figure 2-8.

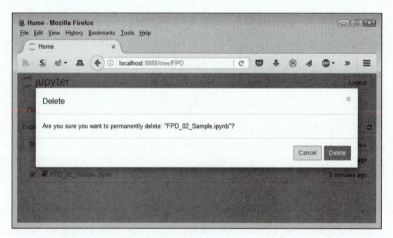

FIGURE 2-8:
Notebook warns you before removing any files from the repository.

3. Click Delete.

The file gets removed from the list.

Importing a notebook

To use the source code from this book, you must import the downloaded files into your repository. The source code comes in an archive file that you extract to a location on your hard drive. The archive contains a list of .ipynb (IPython Notebook) files containing the source code for this book (see the Introduction for details on downloading the source code). The following steps tell how to import these files into your repository:

1. **Click Upload at the top of the page.**

   What you see depends on your browser. In most cases, you see some type of File Upload dialog box that provides access to the files on your hard drive.

2. **Navigate to the directory containing the files that you want to import into Notebook.**

3. **Highlight one or more files to import and click the Open (or other, similar) button to begin the upload process.**

   You see the file added to an upload list, as shown in Figure 2-9. The file isn't part of the repository yet — you've simply selected it for upload.

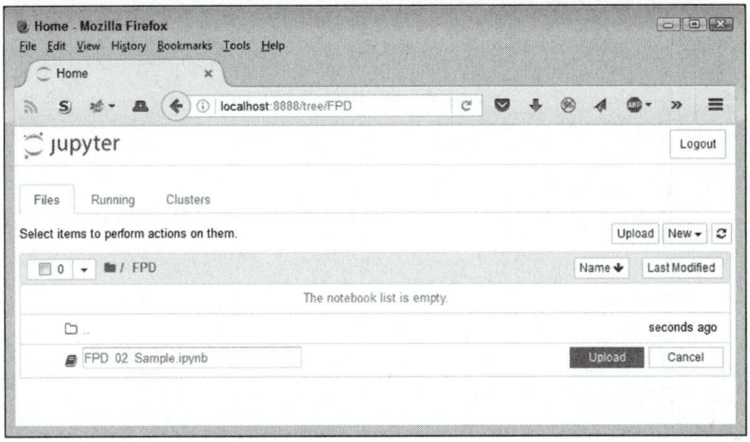

FIGURE 2-9: The files that you want to add to the repository appear as part of an upload list consisting of one or more filenames.

4. **Click Upload.**

   Notebook places the file in the repository so that you can begin using it.

Getting and using datasets

This book uses a number of datasets, all of which appear in the Scikit-learn library. These datasets demonstrate various ways in which you can interact with data, and you use them in the examples to perform a variety of tasks. The following list provides a quick overview of the function used to import each of the datasets into your Python code:

» `load_boston()`: Regression analysis with the Boston house-prices dataset

» `fetch_olivetti_faces()`: Olivetti faces dataset from AT&T

» `make_blobs()`: Generates isotropic Gaussian blobs used for clustering

The technique for loading each of these datasets is the same across examples. The following example shows how to load the Boston house-prices dataset. You can find the code in the FPD_02_Dataset_Load.ipynb notebook.

```
from sklearn.datasets import load_boston
Boston = load_boston()
print(Boston.data.shape)
```

To see how the code works, click Run. The output from the `print` call is (506, 13). You can see the output shown in Figure 2-10.

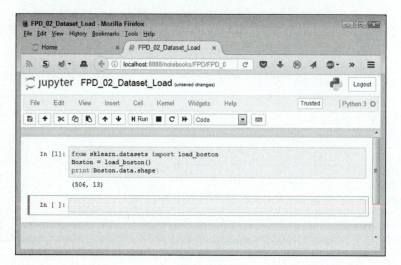

FIGURE 2-10:
The Boston object
contains the
loaded dataset.

TECHNICAL
STUFF

The line `from sklearn.datasets import load_boston` is special because it tells Python to use an external module. In this case, the external module is called `sklearn.datasets`, and Python loads the `load_boston` function from it. After the function is loaded, you can call it from your code, as shown in the next line. You see external modules used quite often in the book, so for now you just need to know that they exist and that you can load them as needed.

Creating a Python Application

Actually, you've already created your first Anaconda application by using the steps in the "Creating a new notebook" section, earlier in this chapter. The `print()` method may not seem like much, but you use it quite often. However, the literate programming approach provided by Anaconda requires a little more knowledge than you currently have. The following sections don't tell you everything about this approach, but they do help you gain an understanding of what literate programming can provide in the way of functionality. However, before you begin, make sure you have the FPD_02_Sample.ipynb file open for use because you need it to explore Notebook.

Understanding cells

If Notebook were a standard IDE, you wouldn't have cells. What you'd have is a document containing a single, contiguous series of statements. To separate various coding elements, you need separate files. Cells are different because each cell is separate. Yes, the results of things you do in previous cells matter, but if a cell is meant to work alone, you can simply go to that cell and run it. To see how this works for yourself, type the following code into the next cell of the FPD_02_Sample file:

```
myVar = 3 + 4
print(myVar)
```

Now click Run (the right-pointing arrow). The code executes, and you see the output, as shown in Figure 2-11. The output is 7, as expected. However, notice the In [1]: entry. This entry tells you that this is the first cell executed during this session. If you want to start a new session (and therefore restart the numbers at 1), you choose Kernel➪Restart (or one of the other restart options).

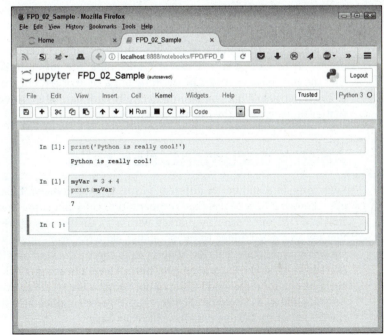

FIGURE 2-11:
Cells execute individually in Notebook.

Note that the first cell also has an In [1]: entry. This entry is still from the previous session. Place your cursor in that cell and click Run. Now the cell contains In [2]:, as shown in Figure 2-12. However, note that the next cell hasn't been selected and still contains the In [1]: entry.

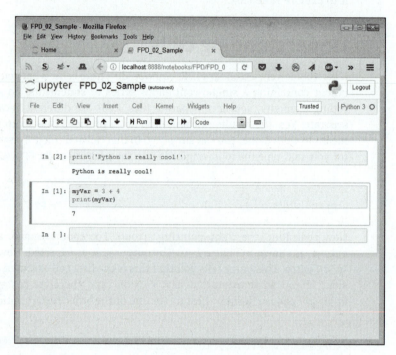

FIGURE 2-12:
Cells can execute in any order in Notebook.

Now place the cursor in the third cell — the one that is currently blank — and type `print("This is myVar: ", myVar)`. Click Run. The output in Figure 2-13 shows that the cells have executed in anything but a rigid order, but that `myVar` is global to the notebook. What you do in other cells with data affects every other cell, no matter in what order the execution takes place.

Adding documentation cells

Cells come in a number of different forms. This book doesn't use them all. However, knowing how to use the documentation cells can come in handy. Select the first cell (the one currently marked with a 2). Choose Insert ➪ Insert Cell Above. You see a new cell added to the notebook. Note the drop-down list that currently shows the word *Code*. This list allows you to choose the kind of cell to create. Select Markdown from the list and type **# This is a level 1 heading**. Click Run (which may seem like an extremely odd thing to do, but give it a try). You see the text change into a heading, as shown in Figure 2-14. However, notice also that the cell lacks the In [x] entry beside it, as the code cells have.

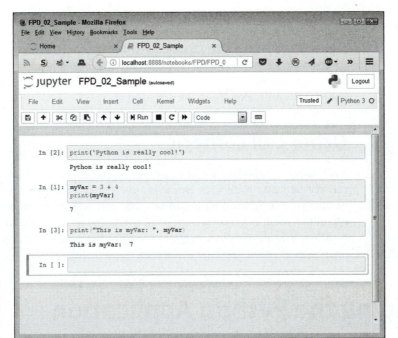

FIGURE 2-13:
Data changes do affect every cell that uses the modified variable.

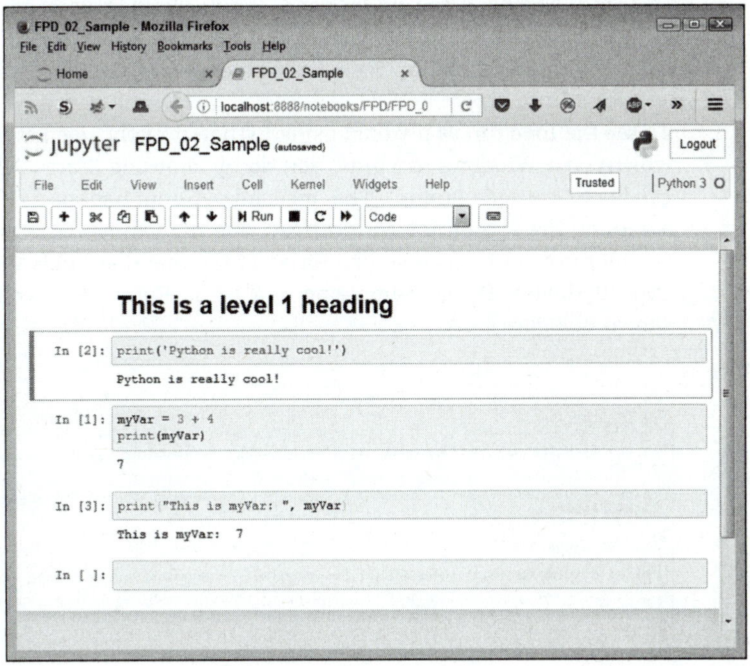

FIGURE 2-14:
Adding headings helps you separate and document your code.

About now, you might be thinking that these special cells act just like HTML pages, and you'd be right. Choose Insert ➪ Insert Cell Below, select Markdown in the drop-down list, and then type **## This is a level 2 heading.** Click Run. As you can see, the number of hashes (#) you add to the text affects the heading level, but the hashes don't show up in the actual heading.

Other cell content

This chapter (and book) doesn't demonstrate all the kinds of cell content that you can see by using Notebook. However, you can add things like graphics to your notebooks, too. When the time comes, you can output (print) your notebook as a report and use it in presentations of all sorts. The literate programming technique is different from what you may have used in the past, but it has definite advantages, as you see in upcoming chapters.

Running the Python Application

The code you create using Notebook is still code and not some mystical unique file that only Notebook can understand. When working with any file, such as the FPD_02_Sample, you can choose File ➪ Download As ➪ Python (.py) to output the Notebook as a Python file. Try it and you end up with FPD_02_Sample.py.

To see the code run as it would using Python directly, open an Anaconda Prompt, which, on a Windows machine, you do by choosing Start ➪ All Programs ➪ Anaconda3 ➪ Anaconda Prompt. The Anaconda Prompt has special features that make accessing the Python interpreter easy. Use the Change Directory (CD) command for your system to change directories to the one that holds the source code file. Type **Python FPD_02_Sample.py** and press Enter. Your code will execute as shown in Figure 2-15.

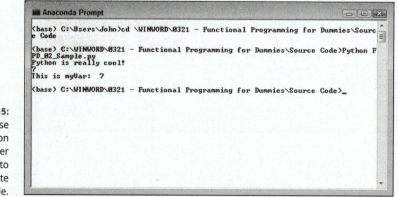

FIGURE 2-15:
You can use the Python interpreter directly to execute your code.

This book doesn't spend much time using this approach because, as you can see, it's harder to use and understand than working with Notebook. However, it's still a perfectly acceptable way to execute your own code.

Understanding the Use of Indentation

As you work through the examples in this book, you see that certain lines are indented. In fact, the examples also provide a fair amount of white space (such as extra lines between lines of code). Python ignores extra lines for the most part, but relies on indentation to show certain coding elements (so the use of indentation is essential). For example, the code associated with a function is indented under that function so that you can easily see where the function begins and ends. The main reason to add extra lines is to provide visual cues about your code, such as the end of a function or the beginning of a new coding element.

The various uses of indentation will become more familiar as go through the examples in the book. However, you should know at the outset why indentation is used and how it gets put in place. To that end, it's time for another example. The following steps help you create a new example that uses indentation to make the relationship between application elements a lot more apparent and easier to figure out later:

1. **Choose New ⇨ Python3.**

Jupyter Notebook creates a new notebook for you. The downloadable source uses the filename FPD_02_Indentation.ipynb, but you can use any name you want.

2. **Type** print("This is a really long line of text that will " +.

You see the text displayed normally onscreen, just as you expect. The plus sign (+) tells Python that there is additional text to display. Adding text from multiple lines together into a single long piece of text is called *concatenation*. You learn more about using this feature later in the book, so you don't need to worry about it now.

3. **Press Enter.**

The insertion point doesn't go back to the beginning of the line, as you might expect. Instead, it ends up directly under the first double quote, as shown in Figure 2-16. This feature is called automatic indention and is one of the features that differentiates a regular text editor from one designed to write code.

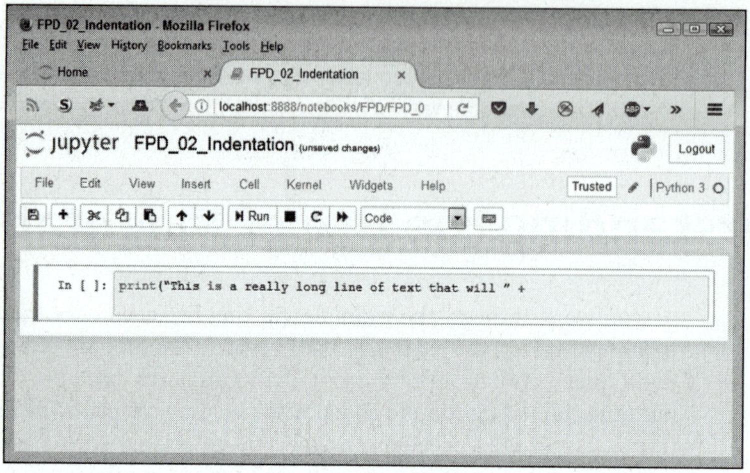

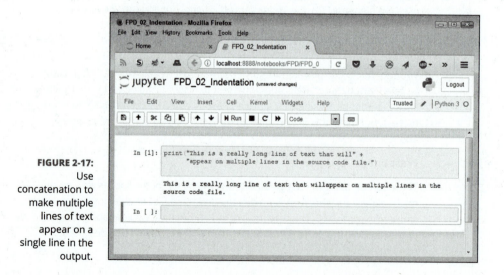

FIGURE 2-16:
The Edit window automatically indents some types of text.

4. **Type** "appear on multiple lines in the source code file.") **and press Enter.**

Notice that the insertion point goes back to the beginning of the line. When Notebook senses that you have reached the end of the code, it automatically outdents the text to its original position.

5. **Click Run.**

You see the output shown in Figure 2-17. Even though the text appears on multiple lines in the source code file, it appears on just one line in the output. The line does break because of the size of the window, but it's actually just one line.

FIGURE 2-17:
Use concatenation to make multiple lines of text appear on a single line in the output.

Adding Comments

People create notes for themselves all the time. When you need to buy groceries, you look through your cabinets, determine what you need, and write it down on a list. When you get to the store, you review your list to remember what you need. Using notes comes in handy for all sorts of needs, such as tracking the course of a conversation between business partners or remembering the essential points of a lecture. Humans need notes to jog their memories. *Comments* in source code are just another form of note. You add comments to the code so that you can remember what task the code performs later. The following sections describe comments in more detail. You can find these examples in the FPD_02_Comments.ipynb file in the downloadable source.

Understanding comments

Computers need some special way to determine that the text you're writing is a comment, not code to execute. Python provides two methods of defining text as a comment and not as code. The first method is the single-line comment. It uses the hash, also called the number sign (#), like this:

```
# This is a comment.
print("Hello from Python!") #This is also a comment.
```

REMEMBER

A single-line comment can appear on a line by itself or after executable code. It appears on only one line. You typically use a single-line comment for short descriptive text, such as an explanation of a particular bit of code. Notebook shows comments in a distinctive color (usually blue) and in italics.

Python doesn't actually support a multiline comment directly, but you can create one using a triple-quoted string. A multiline comment both starts and ends with three double quotes (""") or three single quotes (''') like this:

```
"""

    Application: Comments.py
    Written by: John
    Purpose: Shows how to use comments.
"""
```

REMEMBER

These lines aren't executed. Python won't display an error message when they appear in your code. However, Notebook treats them differently, as shown in Figure 2-18. Note that the actual Python comments, those preceded by a hash (#) in cell 1, don't generate any output. The triple-quote strings, however, do generate output. If you plan to output your notebook as a report, you need to avoid using triple-quoted strings. (Some IDEs, such as IDLE, ignore the triple-quoted strings completely.)

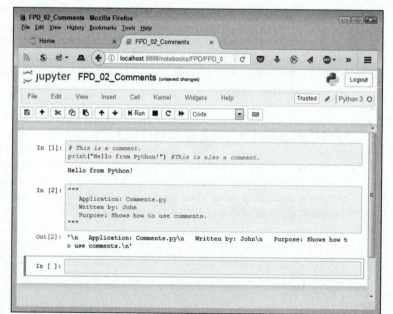

FIGURE 2-18:
Multiline comments do work, but they also provide output.

You typically use multiline comments for longer explanations of who created an application, why it was created, and what tasks it performs. Of course, no hard rules exist for precisely how to use comments. The main goal is to tell the computer precisely what is and isn't a comment so that it doesn't become confused.

Using comments to leave yourself reminders

A lot of people don't really understand comments—they don't quite know what to do with notes in code. Keep in mind that you might write a piece of code today and then not look at it for years. You need notes to jog your memory so that you remember what task the code performs and why you wrote it. Here are some common reasons to use comments in your code:

>> Reminding yourself about what the code does and why you wrote it

>> Telling others how to maintain your code

>> Making your code accessible to other developers

>> Listing ideas for future updates

>> Providing a list of documentation sources you used to write the code

>> Maintaining a list of improvements you've made

You can use comments in a lot of other ways, too, but these are the most common ways. Look at how comments are used in the examples in the book, especially as you get to later chapters where the code becomes more complex. As your code becomes more complex, you need to add more comments and make the comments pertinent to what you need to remember about it.

Using comments to keep code from executing

Developers also sometimes use the commenting feature to keep lines of code from executing (referred to as *commenting out*). You might need to do this to determine whether a line of code is causing your application to fail. As with any other comment, you can use either single-line commenting or multiline commenting. However, when using multiline commenting, you do see the code that isn't executing as part of the output (and it can actually be helpful to see where the code affects the output). Here is an example of both forms of commenting out:

```
# print("This print statement won't print")

"""
    print("This print statement appears as output")
"""
```

Closing Jupyter Notebook

After you have used the File ⇨ Close and Halt command to close each of the notebooks you have open (the individual browser windows), you can simply close the browser window showing the Notebook Home page to end your session. However, the Notebook server (a separate part of Notebook) continues to run in the background. Normally, a Jupyter Notebook window opens when you start Notebook, like the one shown in Figure 2-19. This window remains open until you stop the server. Simply press Ctrl+C to end the server session, and the window will close.

```
Jupyter Notebook
833a2de7b6 (::1) 1.00ms referer=http://localhost:8888/notebooks/FPD/FPD_02_Comme
nts.ipynb
[I 15:28:55.208 NotebookApp] Kernel started: 9b583a81-c264-45af-aa2a-837e64d78e3
f
[I 15:28:56.947 NotebookApp] Adapting to protocol v5.1 for kernel 9b583a81-c264-
45af-aa2a-837e64d78e3f
[I 15:30:56.774 NotebookApp] Saving file at /FPD/FPD_02_Comments.ipynb
[I 15:34:56.751 NotebookApp] Saving file at /FPD/FPD_02_Comments.ipynb
[I 15:36:56.748 NotebookApp] Saving file at /FPD/FPD_02_Comments.ipynb
[I 15:37:04.149 NotebookApp] Starting buffering for 9b583a81-c264-45af-aa2a-837e
64d78e3f:c9b74b7f0a44434383db7f8b65052c0b
[I 15:37:04.494 NotebookApp] Kernel restarted: 9b583a81-c264-45af-aa2a-837e64d78
e3f
[I 15:37:06.326 NotebookApp] Adapting to protocol v5.1 for kernel 9b583a81-c264-
45af-aa2a-837e64d78e3f
[I 15:37:06.327 NotebookApp] Restoring connection for 9b583a81-c264-45af-aa2a-83
7e64d78e3f:c9b74b7f0a44434383db7f8b65052c0b
[I 15:37:06.327 NotebookApp] Replaying 6 buffered messages
[I 15:38:56.748 NotebookApp] Saving file at /FPD/FPD_02_Comments.ipynb
[I 15:42:09.864 NotebookApp] Saving file at /FPD/FPD_02_Comments.ipynb
[I 15:42:24.716 NotebookApp] Starting buffering for 9b583a81-c264-45af-aa2a-837e
64d78e3f:c9b74b7f0a44434383db7f8b65052c0b
[I 15:42:25.065 NotebookApp] Kernel shutdown: 9b583a81-c264-45af-aa2a-837e64d78e
3f
```

FIGURE 2-19: Make sure to close the server window.

TECHNICAL STUFF

Look again at Figure 2-19 to note a number of commands. These commands tell you what the user interface is doing. By monitoring this window, you can determine what might go wrong during a session. Even though you won't use this feature very often, it's a handy trick to know.

Getting Help with the Python Language

You have access to a wealth of Python resources online, and many of them appear in this book in the various chapters. However, the one resource you need to know about immediately is Anaconda Navigator. You start this application by choosing the Anaconda Navigator entry in the Anaconda3 folder. The application requires a few moments to start, so be patient.

REMEMBER

The Home, Environments, and Projects tabs are all about working with the Anaconda tools and utilities. The Learning tab, shown in Figure 2-20, is different because it gives you standardized access to Python-related documentation, training, videos, and webinars. To use any of these resources, simply click the one you want to see or access.

FIGURE 2-20:
Use the Learning tab to get standardized information.

TIP

Note that the page contains more than just Python-specific or Anaconda-specific resources. You also gain access to information about common Python resources, such as the SciPy library.

The Community tab, shown in Figure 2-21, provides access to events, forums, and social entities. Some of this content changes over time, especially the events. To get a quick overview of an entry, hover the mouse over it. Reading an overview is especially helpful when deciding whether you want to learn more about events.

Forums differ from social media by the level of formality and the mode of access. For example, the Stack Overflow allows you to ask Python-related questions, and Twitter allows you to rave about your latest programming feat.

FIGURE 2-21:
Use the Community tab to discover interactive information resources.

Chapter **3**

Getting and Using Haskell

The first sections of this chapter discuss the goals behind the Haskell installation for this book, help you obtain a copy of Haskell, and then show you how to install Haskell on any one of the three supported book platforms: Linux, Mac, and Windows. Overall, this chapter focuses on providing you with the simplest possible installation so that you can clearly see how the functional programming paradigm works. You may eventually find that you need a different installation to meet specific needs or tool requirements.

After you have Haskell installed, you perform some simple coding tasks using it. The main purpose of writing this code is to verify that your copy of Haskell is working properly, but it also helps familiarize you with Haskell just a little. A second example helps you become familiar with using Haskell libraries, which is important when viewing the examples in this book.

REMEMBER

The final section of the chapter helps you locate some Haskell resources. This book doesn't provide you with a solid basis for learning how to program in Haskell. Rather, it focuses on the functional programming paradigm, which can rely on Haskell for a pure implementation approach. Consequently, even though the text gives some basic examples, it doesn't provide a complete treatment of the language, and the aforementioned other resources will help you fill in the gaps if you're new to Haskell.

Working with Haskell in This Book

You can encounter many different, and extremely confusing, ways to work with Haskell. All you need to do is perform a Google search and, even if you limit the results to the past year, you find that everyone has a differing opinion as to how to obtain, install, and configure Haskell. In addition, various tools work with Haskell configured in different ways. You also find that different platforms support different options. Haskell is both highly flexible and relatively new, so you have stability issues to consider. This chapter helps you create a Haskell configuration that's easy to work with and allows you to focus on the task at hand, which is to discover the wonders of the functional programming paradigm.

To ensure that the code that you find in this book works well, make sure to use the 8.2.2 version of Haskell. Older versions may lack features or require bug fixes to make the examples work. You also need to verify that you have a compatible installation by using the instructions found in the upcoming "Obtaining and Installing Haskell" section. Haskell provides a number of very flexible installation options that may not be compatible with the example code.

Obtaining and Installing Haskell

You can obtain Haskell for each of the three platforms supported by this book at `https://www.haskell.org/platform/prior.html`. Simply click the icon corresponding to the platform of your choice. The page takes you to the section that corresponds with the platform.

REMEMBER

In all three cases, you want to perform a full installation, rather than a core installation, because the core installation doesn't provide support for some of the packages used in the book. Both Mac and Windows users can use only a 64-bit installation. In addition, unless you have a good reason to do otherwise, Mac users should rely on the installer, rather than use Homebrew Cask. Linux users should rely on the 64-bit installation as well because you obtain better results. Make sure that you have plenty of drive space for your installation. For example, even though the Windows download file is only 269MB, the Haskell Platform folder will consume 2.6GB of drive space after the installation is complete.

TIP

You can encounter a problem when clicking the links on the initial page. If you find that the download won't start, go to `https://downloads.haskell.org/~platform/8.2.2/` instead and choose the particular link for your platform:

- **>> Generic Linux:** haskell-platform-8.2.2-unknown-posix--full-i386.tar.gz

- **>> Specific Linux:** See the installation instructions in the "Installing Haskell on a Linux system" section that follows

- **>> Mac:** Haskell Platform 8.2.2 Full 64bit-signed.pkg

- **>> Windows:** HaskellPlatform-8.2.2-full-x86_64-setup.exe

Haskell supports some Linux distributions directly. If this is the case, you don't need to download a copy of the product. The following sections get you started with the various installations.

USING HASKELL IDEs AND ENVIRONMENTS

You can find a number of IDEs and environments online for Haskell. Many of these options, such as Vim (https://www.vim.org/download.php), neoVim (https://neovim.io/), and Emacs (https://www.gnu.org/software/emacs/download.html), are enhanced text editors. The problem is that the editors provide uneven feature sets for the platforms that they support. In addition, in each case you must perform additional installations to obtain Haskell support. For example, emacs requires the use of haskell-mode (https://github.com/haskell/haskell-mode/wiki). Consequently, you won't find them used in this book.

Likewise, you can find a Jupyter Notebook add-on for Haskell at https://github.com/gibiansky/IHaskell. The add-on works well as long as you have either Mac or supported Linux as your platform. No Windows support exists for this add-on unless you want to create a Linux virtual machine in which to run it. You can read a discussion of the issues surrounding this add-on at https://news.ycombinator.com/item?id=12783913.

Yet another option is a full-blown Integrated Development Environment (IDE), such as Leksah (http://leksah.org/), which is *Haskell* spelled backward with just one *L*, or HyperHaskell (https://github.com/HeinrichApfelmus/hyper-haskell). Most of these IDEs require that you perform a build, and the setups can become horribly complex for the novice developer. Even so, an IDE can give you advanced functionality, such as a debugger. There really isn't a correct option, but the focus of this book is to make things simple.

Installing Haskell on a Linux system

Linux users numerous options from which to choose. If you see instructions for your particular Linux distribution, you may not even need to download Haskell directly. The $ sudo apt-get command may do everything needed. Use this option if possible. Otherwise, rely on the installation tarball for generic Linux. The specific Linux installations are:

» Ubuntu

» Debian

» Linux Mint

» Redhat

» Fedora

» Gentoo

A generic Linux installation assumes that you don't own one of the distributions in the previous list. In this case, make sure that you download the tarball found in the introduction to this section and follow these instructions to install it:

1. **Type** tar xf haskell-platform-8.2.2-unknown-posix--full-i386.tar.gz **and press Enter.**

   The system extracts the required files for you.

2. **Type** sudo ./install-haskell-platform.sh **and press Enter.**

   The system performs the required installation for you. You'll likely see prompts during the installation process, but these prompts vary by system. Simply answer the questions as you proceed to complete the installation.

WARNING

This book won't help you build Haskell from source, and the results are unreliable enough that this approach isn't recommended for the novice developer. If you find that you absolutely must build Haskell from source files, make sure that you rely on the instructions found in the README file provided with the source code, rather than online instructions that may reflect the needs of an older version of Linux.

Installing Haskell on a Mac system

When working with a Mac platform, you need to access a Haskell installer specifically designed for a Mac. This chapter assumes that you don't want to take time or effort to create a custom configuration using source code. The following steps describe how to perform the installation using the graphical installer.

1. **Locate the downloaded copy of Haskell Platform 8.2.2 Full 64bit-signed. pkg on your system.**

   If you use some other version, you may experience problems with the source code and need to make adjustments when working with it.

2. **Double-click the installation file.**

   You see a Haskell Platform 8.2.2 64-bit Setup dialog box.

3. **Click Next.**

   The wizard displays a licensing agreement. Be sure to read the licensing agreement so that you know the terms of usage.

4. **Click I Agree if you agree to the licensing agreement.**

   The setup wizard asks where you want to install your copy of Haskell. This book assumes that you use the default installation location.

5. **Click Next.**

   You see a dialog box asking which features to install. This book assumes that you install all the default features.

6. **Click Next.**

   You see a new dialog box appear that asks where to install the Haskell Stack. Use the default installation location to ensure that your setup works correctly.

7. **Click Next.**

   The setup wizard asks you which features to install. You must install all of them.

8. **Click Install.**

   You see the Haskell Stack Setup wizard complete.

9. **Click Close.**

   You see the Haskell Platform wizard progress indicator move. At some point, the installation completes.

10. **Click Next.**

    You see a completion dialog box.

11. **Click Finish.**

    Haskell is now ready for use on your system.

Installing Haskell on a Windows system

When working with a Windows platform, you need access to a Haskell installer specifically designed for Windows. The following steps assume that you've down-loaded the required file, as described in the introduction to this section.

1. **Locate the downloaded copy of HaskellPlatform-8.2.2-full-x86_64-setup. exe on your system.**

   If you use some other version, you may experience problems with the source code and need to make adjustments when working with it.

2. **Double-click the installation file.**

   (You may see an Open File – Security Warning dialog box that asks whether you want to run this file. Click Run if you see this dialog box pop up.) You see an Haskell Platform 8.2.2 64-bit Setup dialog box.

3. **Click Next.**

   The wizard displays a licensing agreement. Be sure to read through the licensing agreement so that you know the terms of usage.

4. **Click I Agree if you agree to the licensing agreement.**

   The setup wizard asks where you want to install your copy of Haskell, as shown in Figure 3-1. This book assumes that you use the default installation location, but you can enter a different one.

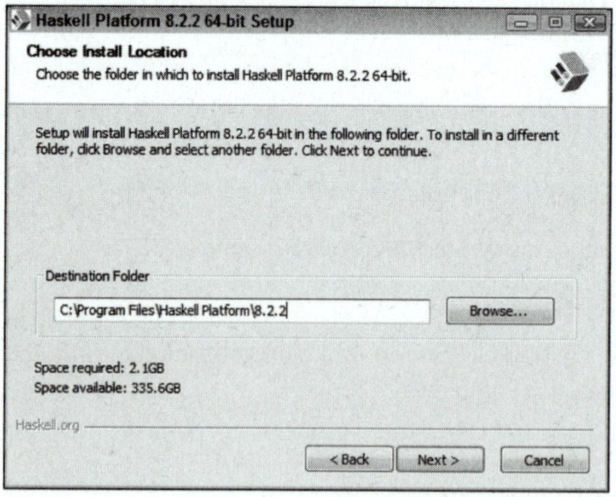

FIGURE 3-1: Specify a Haskell installation location.

5. **Optionally provide an installation location and then click Next.**

You see a dialog box asking which features to install. This book assumes that you install all the default features, as shown in Figure 3-2. Note especially the Update System Settings option. You must ensure that this option is selected to obtain proper functioning of the Haskell features.

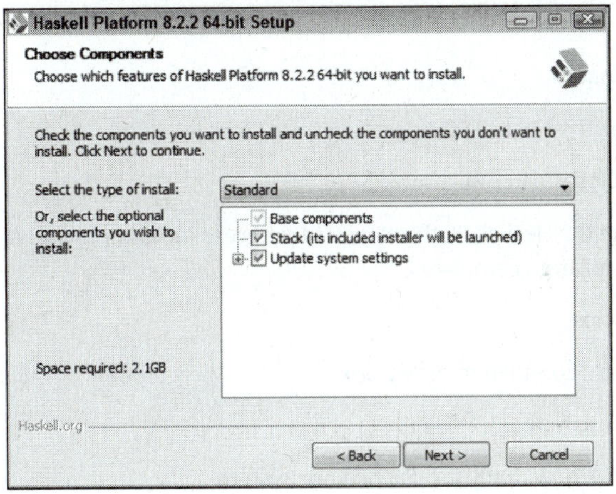

FIGURE 3-2: Choose which Haskell features to install.

6. **Choose the features you want to use and click Next.**

The setup wizard asks you which Start menu folder to use, as shown in Figure 3-3. The book assumes that you use the default Start menu folder, but you can enter a name that you choose.

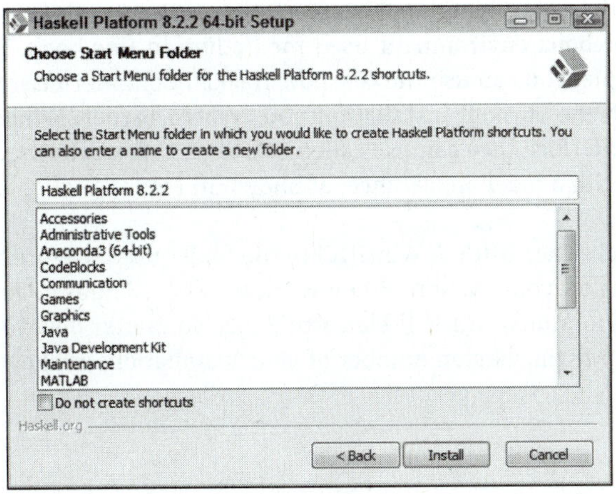

FIGURE 3-3: Type a Start menu folder name, if desired.

7. Optionally type a new Start menu folder name and click Install.

You see a new dialog box appear that asks where to install the Haskell Stack. Use the default installation location unless you need to change it for a specific reason, such as using a local folder rather than a roaming folder.

8. Optionally type a new location and click Next.

The setup wizard asks you which features to install. You must install all of them.

9. Click Install.

You see the Haskell Stack Setup wizard complete.

10. Click Close.

You see the Haskell Platform wizard progress indicator move. At some point, the installation completes.

11. Click Next.

You see a completion dialog box.

12. Click Finish.

Haskell is now ready for use on your system.

Testing the Haskell Installation

As explained in the "Using Haskell IDEs and Environments" sidebar, you have access to a considerable number of environments for working with Haskell. In fact, if you're using Linux or Mac platforms, you can rely on an add-in for the Jupyter Notebook environment used for Python in this book. However, to make things simple, you can use the Glasgow Haskell Compiler interpreter (GHCi) that comes with the Haskell installation you created earlier. Windows users have a graphical interface they can use called WinGHCi that works precisely the same as GHCi, but with a nicer appearance, as shown in Figure 3-4.

You can find either GHCi or WinGHCi in the folder used to store the Haskell application icons on your system. When working with Windows, you find this file at Start ⇨ All Programs ⇨ Haskell Platform 8.2.2. No matter how you open the interpreter, you see the version number of your installation, as shown in Figure 3-4.

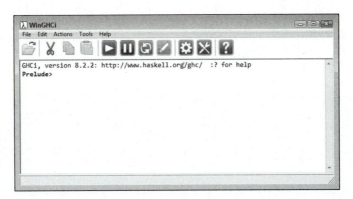

FIGURE 3-4:
The WinGHCi interface offers a nice appearance and is easy to use.

REMEMBER

The interpreter can provide you with a great deal of information about Haskell, and simply looking at what's available can be fun. The commands all start with a colon, including the help commands. So to start the process, you type **:?** and press Enter. Figure 3-5 shows typical results.

FIGURE 3-5:
Make sure to precede all help commands with a colon (:) in the interpreter.

As you look through the list, you see that all commands begin with a colon. For example, to exit the Haskell interpreter, you type **:quit** and press Enter.

Playing with Haskell is the best way to learn it. Type **"Haskell is fun!"** and press Enter. You see the string repeated onscreen, as shown in Figure 3-6. All Haskell has done is evaluate the string you provided.

As a next step, try creating a variable by typing **x = "Haskell is really fun!"** and pressing Enter. This time, Haskell doesn't interpret the information but simply places the string in x. To see the string, you can use the putStrLn function. Type **putStrLn x** and press Enter. Figure 3-7 shows what you should see. At this point, you know that the Haskell installation works.

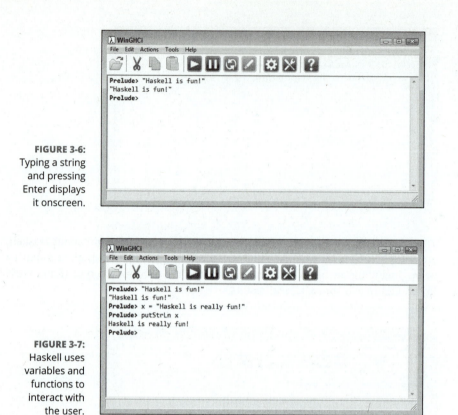

FIGURE 3-6:
Typing a string and pressing Enter displays it onscreen.

FIGURE 3-7:
Haskell uses variables and functions to interact with the user.

Compiling a Haskell Application

Even though you'll perform most tasks in this book using the interpreter, you can also load modules and interpret them. In fact, this is how you use the download-able source: You load it into the interpreter and then execute it. To see how this works, create a text file on your system called Simple.hs. You must use a pure text editor (one that doesn't include any formatting in the output file), such as Notepad or TextEdit. Type the following code into the file and save it on disk:

```
main = putStrLn out
    where
        out = "5! = " ++ show result
        result = fac 5

fac 0 = 1
fac n = n * fac (n - 1)
```

TECHNICAL STUFF

This code actually demonstrates a number of Haskell features, but you don't need to fully understand all of them now. To compile a Haskell application, you must have a main function, which consists of a single statement, which in this case is `putStrLn out`. The variable `out` is defined as part of the `where` clause as the concatenation of a string, `"5! = "`, and an integer, `result`, that you output using the `show` function. Notice the use of indentation. You must indent the code for it to compile correctly, which is actually the same use of indentation as found in Python.

The code calculates the result by using the `fac` (factorial) function that appears below the `main` function. As you can see, Haskell makes it easy to use recursion. The first line defines the stopping point. When the input is equal to `0`, the function outputs a value of `1`. Otherwise, the second line is used to call `fac` recursively, with each succeeding call reducing the value of `n` by `1` until `n` reaches `0`.

After you save the file, you can open GHCi or WinGHCi to experiment with the application. The following steps provide the means to load, test, and compile the application:

1. **Type** :cd *<Source Code Directory>* **and press Enter.**

   Supply the location of the source code on your system. The location of your source code will likely differ from mine.

2. **Type** :load Simple.hs **and press Enter.**

   Notice that the prompt changes to *Main>, as shown in Figure 3-8. If you're using WinGHCi, you can also use the File ⇨ Load menu command to accomplish this task.

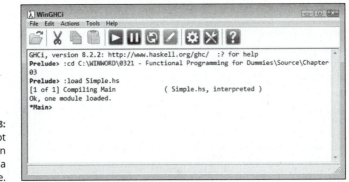

FIGURE 3-8:
The prompt changes when you load a source file.

3. Type :main and press Enter.

You see the output of the application as shown in Figure 3-9. When working with WinGHCi, you can also use the Actions ➪ Run "main" command or you can click the red button with the right-pointing arrow on the toolbar.

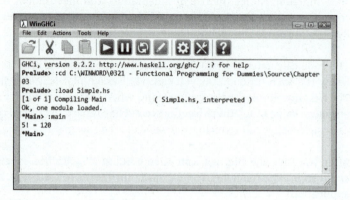

FIGURE 3-9:
Executing the
`main` function
shows what
the application
file can do.

4. Type :! ghc --make "Simple.hs" and press Enter.

The interpreter now compiles the application, as shown in Figure 3-10. You see a new executable created in the source code directory. When working with WinGHCi, you can also use the Tools ➪ GHC Compiler menu command to perform this task. You can now execute the application at the command prompt and get the same results as you did in the interpreter.

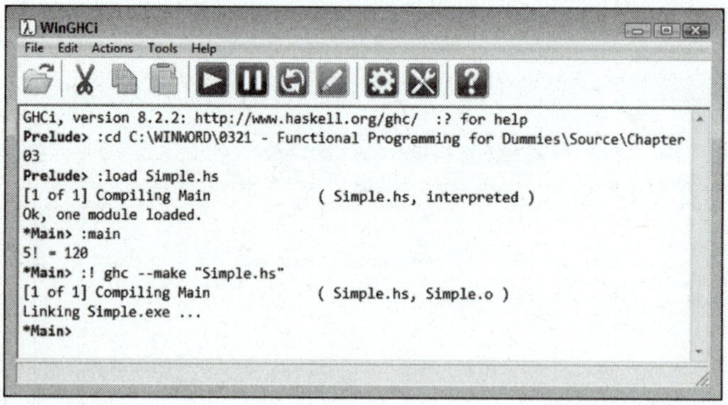

FIGURE 3-10:
Compiling
the loaded
module creates
an executable
on disk.

5. Type :module and press Enter.

This act unloads all the existing modules. Notice that the prompt changes back to Prelude>. You can also perform this task using the Actions ➪ Clear Modules menu command.

6. **Type** :quit **and press Enter.**

The interpreter closes. You're done working with Haskell for now.

These steps show just a small sample of the kinds of tasks you can perform using GHCi. As the book progresses, you see how to perform more tasks, but this is a good start on discovering what Haskell can do for you.

Using Haskell Libraries

Haskell has a huge library support base in which you can find all sorts of useful functions. Using library code is a time saver because libraries usually contain well-constructed and debugged code. The `import` function allows you to use external code. The following steps take you through a simple library usage example:

1. **Open GHCi, if necessary.**

2. **Type** import Data.Char **and press Enter.**

Note that the prompt changes to Prelude Data.Char> to show that the import is successful. The `Data.Char` library contains functions for working with the `Char` data type. You can see a listing of these functions at `http://hackage.haskell.org/package/base-4.11.1.0/docs/Data-Char.html`. In this case, the example uses the `ord` function to convert a character to its ASCII numeric representation.

3. **Type** ord('a') **and press Enter.**

You see the output value of 97.

The "Getting and using datasets" section of Chapter 2 discusses how to obtain a dataset for use with Python. You can obtain these same datasets for Haskell, but first you need to perform a few tasks. The following steps will work for any platform if you have installed Haskell using the procedure in the earlier part of this chapter.

1. **Open a command prompt or Terminal window with administrator privileges.**

2. **Type** cabal update **and press Enter.**

You see the update process start. The cabal utility provides the means to perform updates in Haskell. The first thing you want to do is ensure that your copy of cabal is up to date.

3. **Type** cabal install Datasets **and press Enter.**

   You see a rather long list of download, install, and configure sequences. All these steps install the Datasets module documented at `https://hackage.haskell.org/package/datasets-0.2.5/docs/Numeric-Datasets.html` onto your system.

4. **Type** cabal list Datasets **and press Enter.**

   The cabal utility outputs the installed status of Datasets, along with other information. If you see that Datasets isn't installed, try the installation again by typing **cabal install Datasets --force-reinstalls** and pressing Enter instead.

Chapter 2 uses the Boston Housing dataset as a test, so this chapter will do the same. The following steps show how to load a copy of the Boston Housing dataset in Haskell.

1. **Open GHCi or WinGHCi.**

2. **Type** import Numeric.Datasets (getDataset) **and press Enter.**

   Notice that the prompt changes. In fact, it will change each time you load a new package. The step loads the `getDataset` function, which you need to load the Boston Housing dataset into memory.

3. **Type** import Numeric.Datasets.BostonHousing (bostonHousing) **and press Enter.**

   The `BostonHousing` package loads as `bostonHousing`. Loading the package doesn't load the dataset. It provides support for the dataset, but you still need to load the data.

4. **Type** bh <- getDataset bostonHousing **and press Enter.**

   This step loads the Boston Housing dataset into memory as the object bh. You can now access the data.

5. **Type** print (length bh) **and press Enter.**

   You see an output of 506, which matches the length of the dataset in Chapter 2.

Getting Help with the Haskell Language

The documentation that the wizard installs as part of your Haskell setup is the first place you should look when you have questions. There are three separate files for answering questions about: GHC, GHC flags, and the Haskell libraries.

In addition, you see a link for HackageDB, which is the Haskell Software Repository where you get packages such as Datasets used in the "Using Haskell Libraries" section of this chapter. All these resources help you see the wealth of functionality that Haskell provides.

Tutorials make learning any language a lot easier. Fortunately, the Haskell community has created many tutorials that take different approaches to learning the language. You can see a listing of these tutorials at `https://wiki.haskell.org/Tutorials`.

No matter how adept you might be, documentation and tutorials won't be enough to solve every problem. With this in mind, you need access to the Haskell community. You can find many different groups online, each with people who are willing to answer questions. However, one of the better places to look for help is StackOverflow at `https://stackoverflow.com/search?q=haskell`.

2

Starting Functional Programming Tasks

Chapter **4**

Defining the Functional Difference

A s described in Chapter 1 and explored in Chapters 2 and 3, using the functional programming paradigm entails an approach to problems that differs from the paradigms that languages have relied on in the past. For one thing, the functional programming paradigm doesn't tie you to thinking about a problem as a machine would; instead, you use a mathematical approach that doesn't really care about how the machine solves the problem. As a result, you focus on the problem description rather than the solution. The difference means that you use *declarations* —formal or explicit statements describing the problem — instead of procedures — step-by-step problem solutions.

REMEMBER

To make the functional paradigm work, the code must manage data differently than when using other paradigms. The fact that functions can occur in any order and at any time (allowing for parallel execution, among other things) means that functional languages can't allow mutable variables that maintain any sort of state or provide side effects. These limitations force developers to use better coding practices. After all, the use of side effects in coding is really a type of shortcut that can make the code harder to understand and manage, besides being far more prone to bugs and other reliability issues.

This chapter provides examples in both Haskell and Python to demonstrate the use of functions. You see extremely simple uses of functions in Chapters 2 and 3, but this chapter helps move you to the next level.

Comparing Declarations to Procedures

The term *declaration* has a number of meanings in computer science, and different people use the term in different ways at different times. For example, in the context of a language such as C, a declaration is a language construct that defines the properties associated with an identifier. You see declarations used for defining all sorts of language constructs, such as types and enumerations. However, that's not how this book uses the term *declaration*. When making a declaration in this book, you're telling the underlying language to do something. For example, consider the following statement:

1. Make me a cup of tea!

The statement tells simply what to do, not how to do it. The declaration leaves the execution of the task to the party receiving it and infers that the party knows how to complete the task without additional aid. Most important, a declaration enables someone to perform the required task in multiple ways without ever changing the declaration. However, when using a procedure named MakeMeTea (the identifier associated with the procedure), you might use the following sequence instead:

1. Go to the kitchen.
2. Get out the teapot.
3. Add water to the teapot.
4. Bring the pot to a boil.
5. Get out a teacup.
6. Place a teabag in the teacup.
7. Pour hot water over the teabag and let steep for five minutes.
8. Remove the teabag from the cup.
9. Bring me the tea.

REMEMBER

A *procedure* details what to do, when to do it, and how to do it. Nothing is left to chance and no knowledge is assumed on the part of the recipient. The steps appear in a specific order, and performing a step out of order will cause problems. For example, imagine pouring the hot water over the teabag before placing the teabag in the cup. Procedures are often error prone and inflexible, but they do allow for precise control over the execution of a task. Even though making a declaration might seem to be superior to a procedure, using procedures does have advantages that you must consider when designing an application.

Declarations do suffer from another sort of inflexibility, however, in that they don't allow for interpretation. When making a declarative statement ("Make me a cup of tea!"), you can be sure that the recipient will bring a cup of tea and not a cup of coffee instead. However, when creating a procedure, you can add conditions that rely on state to affect output. For example, you might add a step to the procedure that checks the time of day. If it's evening, the recipient might return coffee instead of tea, knowing that the requestor always drinks coffee in the evening based on the steps in the procedure. A procedure therefore offers flexibility in its capability to interpret conditions based on state and provide an alternative output.

Declarations are quite strict with regard to input. The example declaration says that a *cup* of tea is needed, not a pot or a mug of tea. The MakeMeTea procedure, however, can adapt to allow variable inputs, which further changes its behavior. You can allow two inputs, one called size and the other beverage. The size input can default to cup and the beverage input can default to tea, but you can still change the procedure's behavior by providing either or both inputs. The identifier, MakeMeTea, doesn't indicate anything other than the procedure's name. You can just as easily call it MyBeverageMaker.

One of the hardest issues in moving from imperative languages to functional languages is the concept of declaration. For a given input, a functional language will produce the same output and won't modify or use application state in any way. A declaration always serves a specific purpose and only that purpose.

The second hardest issue is the loss of control. The language decides how to perform tasks, not the developer. Yet, you sometimes see functional code where the developer tries to write it as a procedure, usually producing a less-than-desirable result (when the code runs at all).

Understanding How Data Works

Data is a representation of something — perhaps a value. However, it can just as easily represent a real-world object. The data itself is always abstract, and existing computer technology represents it as a number. Even a character is a number: The letter *A* is actually represented as the number 65. The letter is a value, and the number is the representation of that value: the data. The following sections discuss data with regard to how it functions within the functional programming paradigm.

Working with immutable data

Being able to change the content of a variable is problematic in many languages. The memory location used by the variable is important. If the data in a particular memory location changes, the value of the variable pointing to that memory location changes as well. The concept of immutable data requires that specific memory locations remain untainted. All Haskell data is immutable.

Python data, on the other hand, isn't immutable in all cases. The "Passing by reference versus by value" section that appears later in the chapter gives you an example of this issue. When working with Python code, you can rely on the id function to help you determine when changes have occurred to variables. For example, in the following code, the output of the comparison between id(x) and oldID will be false.

```
x = 1
oldID = id(x)
x = x + 1
id(x) == oldID
```

Every scenario has some caveats, and doing this with Python does as well. The id of a variable is always guaranteed unique except in certain circumstances:

>> One variable goes out of scope and another is created in the same location.

>> The application is using multiprocessing and the two variables exist on different processors.

>> The interpreter in use doesn't follow the CPython approach to handling variables.

When working with other languages, you need to consider whether the data supported by that language is actually immutable and what set of events occurs when code tries to modify that data. In Haskell, modifications aren't possible, and in Python, you can detect changes, but not all languages support the functionality required to ensure that immutability is maintained.

Considering the role of state

Application *state* is a condition that occurs when the application performs tasks that modify global data. An application doesn't have state when using functional programming. The lack of state has the positive effect of ensuring that any call to a function will produce the same results for a given input every time, regardless of when the application calls the function. However, the lack of state has a negative effect as well: The application now has no memory. When you think about state, think about the capability to remember what occurred in the past, which, in the case of an application, is stored as global data.

Eliminating side effects

Previous discussions of procedures and declarations (as represented by functions) have left out an important fact. Procedures can't return a value. The first section of the chapter, "Comparing Declarations to Procedures," presents a procedure that seems to provide the same result as the associated declaration, but the two aren't the same. The declaration "Make me a cup of tea!" has only one output: the cup of tea. The procedure has a *side effect* instead of a value. After making a cup of tea, the procedure indicates that the recipient of the request should take the cup of tea to the requestor. However, the procedure must successfully conclude for this event to occur. The procedure isn't returning the tea; the recipient of the request is performing that task. Consequently, the procedure isn't returning a value.

Side effects also occur in data. When you pass a variable to a function, the expectation in functional programming is that the variable's data will remain untouched — immutable. A side effect occurs when the function modifies the variable data so that upon return from the function call, the variable changes in some manner.

Seeing a Function in Haskell

Haskell is all about functions, so, unsurprisingly, it supports a lot of function types. This chapter doesn't overwhelm you with a complete listing of all the function types (see Chapter 5, for example, to discover lambda functions), but it does demonstrate two of the more important function types (non-curried and curried) in the following sections.

Using non-curried functions

You can look at non-curried functions as Haskell's form of the standard function found in other languages. The next section explains the issue of currying, but for now, think of standard functions as a stepping-stone to them. To create a standard function, you provide a function description like this one:

```
add (x, y) = x + y
```

This function likely looks similar to functions you create in other languages. To use this function, you simply type something like **add (1, 2)** and press Enter. Figure 4-1 shows the result.

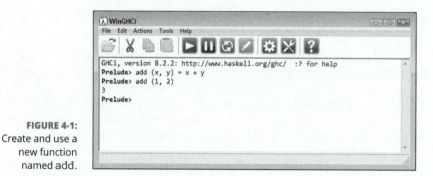

FIGURE 4-1:
Create and use a new function named add.

Functions can act as the basis for other functions. Incrementing a number is really just a special form of addition. Consequently, you can create the inc function shown here:

```
inc (x) = add (x, 1)
```

As you can see, add is the basis for inc. Using inc is as simple as typing something like **inc 5** and pressing Enter. Note that the parentheses are optional, but you could also type **inc (5)** and press Enter. Figure 4-2 shows the result.

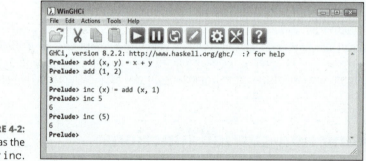

FIGURE 4-2:
Use add as the basis for inc.

Using curried functions

Currying in Haskell is the process of transforming a function that takes multiple arguments into a function that takes just one argument and returns another function when additional arguments are required. The examples in the previous section act as a good basis for seeing how currying works in contrast to non-curried functions. Begin by opening a new window and creating a new version of add, as shown here:

```
add x y = x + y
```

The difference is subtle, but important. Notice that the arguments don't appear in parentheses and have no comma between them. The function content still appears the same, however. To use this function, you simply type something like **add 1 2** and press Enter. Figure 4-3 shows the result.

FIGURE 4-3:
The curried form
of add uses no
parentheses.

You don't actually see the true effect of currying, though, until you create the inc function. The inc function really does look different, and the effects are even more significant when function complexity increases:

```
inc = add 1
```

This form of the inc function is shorter and actually a bit easier to read. It works the same way as the non-curried version. Simply type something like **inc 5** and press Enter to see the result shown in Figure 4-4.

FIGURE 4-4:
Currying makes
creating new
functions easier.

Interestingly enough, you can convert between curried and non-curried versions of a function as needed using the built-in curry and uncurry functions. Try it with add by typing **uadd = uncurry add** and pressing Enter. To prove to yourself that uadd is indeed the non-curried form of add, type **uadd 1 2** and press Enter. You see the error shown in Figure 4-5.

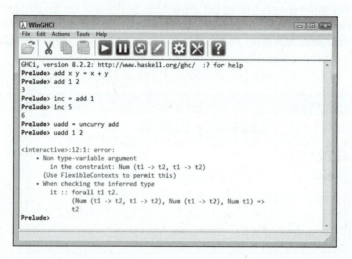

FIGURE 4-5:
The uadd function really is the non-curried form of add.

TECHNICAL STUFF

You can use curried functions in some places where non-curried functions won't work. The map function is one of these situations. (Don't worry about the precise usage of the map function for now; you see it demonstrated in Chapter 6.) The following code adds a value of 1 to each of the members of the list.

```
map (add 1) [1, 2, 3]
```

The output is [2,3,4] as expected. Trying to perform the same task using uadd results in an error, as shown in Figure 4-6.

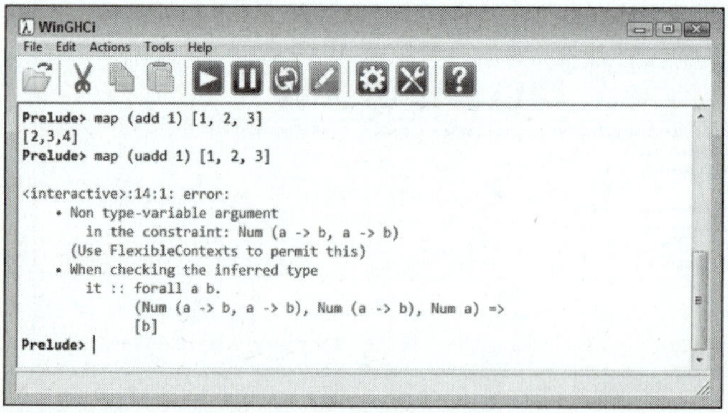

FIGURE 4-6:
Curried functions add essential flexibility to Haskell.

Seeing a Function in Python

Functions in Python look much like functions in other languages. The following sections show how to create and use Python functions, as well as provide a warning about using them in the wrong way. You can compare this section with the previous section to see the differences between pure and impure function use. (The "Defining Functional Programming" section of Chapter 1 describes the difference between pure and impure approaches to functional programming.)

Creating and using a Python function

Python relies on the `def` keyword to define a function. For example, to create a function that adds two numbers together, you can use the following code:

```
def add(x, y):
    return x + y
```

To use this function, you can type something like **add(1, 2)**. Figure 4-7 shows the output of this code when you run it in Notebook.

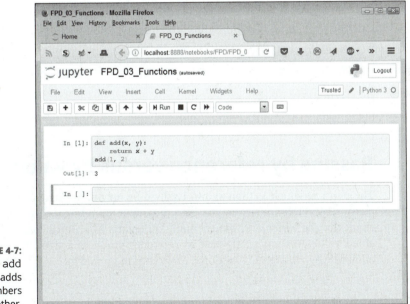

FIGURE 4-7:
The add
function adds
two numbers
together.

As with Haskell, you can use Python functions as the basis for defining other functions. For example, here is the Python version of inc:

```
def inc(x):
    return add(x, 1)
```

The inc function simply adds 1 to the value of any number. To use it, you might type something like **inc(5)** and then run the code, as shown in Figure 4-8, using Notebook.

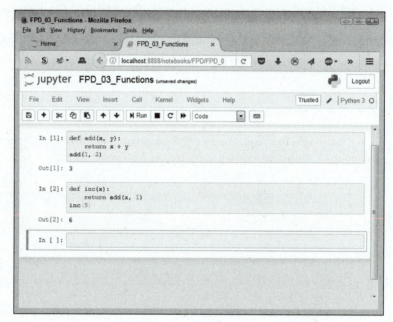

FIGURE 4-8:
You can build functions using other functions as needed.

Passing by reference versus by value

The point at which Python shows itself to be an impure language is the use of passing by reference. When you pass a variable by reference, it means that any change to the variable within the function results in a global change to the variable's value. In short, using pass by reference produces a side effect, which isn't allowed when using the functional programming paradigm.

Normally, you can write functions in Python that don't cause the passing by reference problem. For example, the following code doesn't modify x, even though you might expect it to:

```
def DoChange(x, y):
    x = x.__add__(y)
    return x
x = 1
print(x)
print(DoChange(x, 2))
print(x)
```

The value of x outside the function remains unchanged. However, you need to exercise care when creating functions using some objects and built-in methods. For example, the following code will modify the output:

```
def DoChange(aList):
    aList.append(4)
    return aList
aList = [1, 2, 3]
print(aList)
print(DoChange(aList))
print(aList)
```

REMEMBER

The appended version will become permanent in this case because the built-in function, append, performs the modification. To avoid this problem, you must create a new variable within the function, change its value, and then return the new variable, as shown in the following code:

```
def DoChange(aList):
    newList = aList.copy()
    newList.append(4)
    return newList
aList = [1, 2, 3]
print(aList)
print(DoChange(aList))
print(aList)
```

Figure 4-9 shows the results. In the first case, you see the changed list, but the second case keeps the list intact.

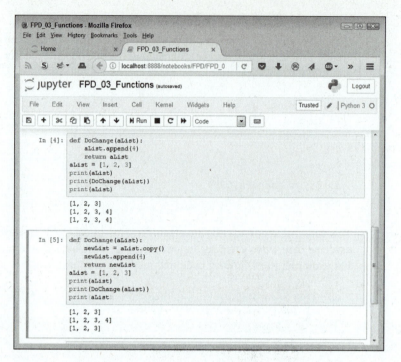

FIGURE 4-9:
Use objects and built-in functions with care to avoid side effects.

TIP

Whether you encounter a problem with particular Python objects or not depends on their mutability. An `int` isn't mutable, so you don't need to worry about having problems with functions changing its value. On the other hand, a `list` is mutable, which is the source of the problems with the examples that use a `list` in this section. The article at https://medium.com/@meghamohan/mutable-and-immutable-side-of-python-c2145cf72747 offers insights into the mutability of various Python objects.

Chapter **5**

Understanding the Role of Lambda Calculus

Mention the word *calculus* and some people automatically assume that the topic is hard or difficult to understand. Add a Greek letter, such as λ (lambda), in front of it and it must be so terribly hard that only geniuses need apply. Of course, using correct terminology is important when discussing a topic for which confusion reigns. The truth is, though, that you've probably used lambda calculus at some point if you've worked with other languages that support first-class functions such as C or JavaScript. Often, the creators of these languages make things simple by using more approachable terms. This chapter helps you through some of the terms involved in lambda calculus while also helping you understand the use of it. The big takeaway from this chapter should be that lambda calculus really isn't hard; you've likely seen it in a number of places before.

REMEMBER

The focus of this chapter is to demonstrate how you can use lambda calculus to solve math problems within an application that relies on the functional programming paradigm. In many cases, the examples look astonishingly simple, and they truly are. When you understand the rules for using lambda calculus, you begin to use it to perform one of three operations — also represented by Greek letters: α (alpha), β (beta), and η (eta). Yes, that's right; you need only to think about three operations, so the task should already be looking easier.

The final section of this chapter shows you how to create and use functions that rely on lambda calculus in the target languages for this book. However, no matter what programming language you use, you can find examples on how to create and use lambda functions (as long as the language supports first-class functions). That's because lambda calculus is so incredibly useful and makes performing programming tasks significantly easier rather than harder, as you might initially expect.

Considering the Origins of Lambda Calculus

Alonzo Church originally created lambda calculus in the 1930s, which is before the time that computers were available. Lambda calculus explores the theoretical basis for what it means to compute. Alonzo Church worked with people like Haskell Curry, Kurt Gödel, Emil Post, and Alan Turing to create a definition for algorithms. The topic is far more familiar today, but imagine that you're one of the pioneers who are trying to understand the very concepts used to make math doable in an automated way. Each person involved in defining what it means to compute approached it in a different manner:

>> **Alonzo Church:** λ-calculus (the topic of this book)

>> **Haskell Curry:** Combinatory logic (see https://wiki.haskell.org/Combinatory_logic for details)

>> **Kurt Gödel:** μ-recursive functions (see http://www.cs.swan.ac.uk/cie06/files/d129/cie-beam.pdf and http://ebooks.bharathuniv.ac.in/gdlc1/gdlc1/Engineering Merged Library v3.0/GDLC/m-Recursive_Functions (5679)/m-Recursive_Functions - GDLC.pdf for details)

>> **Emil Post:** Post-canonical system, also called a rewrite system (see https://www.revolvy.com/main/index.php?s=Post canonical system&nojs=1 and https://esolangs.org/wiki/Post_canonical_system for details)

>> **Alan Turing:** Turing machines (see http://www.alanturing.net/turing_archive/pages/reference articles/what is a turing machine.html for details)

REMEMBER

Even through each approach is different, Church and others noted certain equivalences between each of the systems, which isn't a coincidence. In addition, each system helped further define the others and overcome certain obstacles that each system presents. Precisely who invented what is often a matter for debate because these people also worked with other scientists, such as John von Neumann.

It shouldn't surprise you to know that some of these people actually attended school together at Princeton (along with Albert Einstein). The history of the early years of computing (when modern computers weren't even theories yet) is fascinating, and you can read more at `https://www.princeton.edu/turing/alan/history-of-computing-at-p/`.

TECHNICAL STUFF

Church's motivation in creating lambda calculus was to prove that Hilbert's Entscheidungsproblem, or decision problem (see `https://www.quora.com/How-can-I-explain-Entscheidungs-problem-in-a-few-sentences-to-people-without-confusing-people` for details) wasn't solvable Peano arithmetic (see `http://mathworld.wolfram.com/PeanoArithmetic.html` for details). However, in trying to prove something quite specific, Church created a way of looking at math generally that is still in use today.

The goals of lambda calculus, as Church saw it, are to study the interaction of functional abstraction and function application from an abstract, purely mathematical perspective. *Functional abstraction* begins by breaking a particular problem into a series of steps. Of course, these breaks aren't arbitrary; you must create breaks that make sense. The abstraction continues by mapping each of these steps to a function. If a step can't be mapped to a function, then it isn't a useful step — perhaps the break has come in the wrong place. *Function application* is the act of applying the function to an argument and obtaining a value as output. It's essential to understand the tenets of lambda calculus as set out initially by Church to understand where programming languages stand on the topic today:

>> **Lambda calculus uses only functions — no other data or other types (no strings, integers, Booleans, or other types found in programming languages today).** Any other type is encoded as part of a function and therefore, the function is the basis of everything.

>> **Lambda calculus has no state or side effects.** Consequently, you can view lambda calculus in terms of the substitutional model, which is used for biology to describe how a sequence of symbols changes into another set of traits using a particular process.

>> **The order of evaluation is irrelevant.** However, most programming languages do use a particular order to make the process of evaluating functions easier (not to mention reducing the work required to create a compiler or interpreter).

>> **All functions are unary, taking just one argument.** Functions that require multiple arguments require the use of currying. You can read about the use of currying in Haskell in the "Using curried functions" section of Chapter 4.

Understanding the Rules

As mentioned in the introduction to this chapter, you use three different operations to perform tasks using lambda calculus: creating functions to pass as variables; binding a variable to the expression (abstraction); and applying a function to an argument. The following sections describe all three operations that you can view as rules that govern all aspects of working with lambda calculus.

Working with variables

When considering variables in lambda calculus, the variable is a placeholder (in the mathematical sense) and not a container for values (in the programming sense). Any variable, x, y, or z, (or whatever identifier you choose to use) is a lambda term. Variables provide the basis of the *inductive* (the inference of general laws from specific instances) definition of lambda terms. To put this in easier-to-understand terms, if you always leave for work at 7:00 a.m. and are always on time, inductive reasoning says that you will always be on time as long as you leave by 7:00 a.m.

REMEMBER

Induction in math relies on two cases to prove a property. For example, a common proof is a property that holds for all natural numbers. The base (or basis) case makes an assumption using a particular number, usually 0. The inductive case, also called the inductive step, proves that if the property holds for the first natural number (n), it must also hold for the next natural number ($n + 1$).

Variables may be untyped or typed. Typing isn't quite the same in this case because types are for other programming paradigms; the use of typing doesn't actually indicate a kind of data. Rather, it defines how to interpret the lambda calculus. The following sections describe how untyped and typed variables work.

Untyped

The original version of Church's lambda calculus has gone through a number of revisions as the result of input by other mathematicians. The first such revision came as the result of input from Stephen Kleene and J. B. Rosser in 1935 in the form of the Kleene–Rosser paradox. (The article at `https://www.quora.com/What-is-the-Kleene-Rosser-paradox-in-simple-terms` provides a basic description of this issue.) A problem exists in the way that logic worked in the original version of lambda calculus, and Church fixed this problem in a succeeding version by removing restrictions on the kind of input that a function can receive. In other words, a function has no type requirement.

TIP

The advantage of untyped lambda calculus is its greater flexibility; you can do more with it. However, the lack of type also means that untyped lambda calculus is nonterminating, an issue discussed in the "Considering the need for typing" section of the chapter. In some cases, you must use typed lambda calculus to obtain a definitive answer to a problem.

Simply-typed

Church created simply-typed lambda calculus in 1940 to address a number of issues in untyped lambda calculus, the most important of which is an issue of paradoxes where β-reduction can't terminate. In addition, the use of simple typing provides a means for strongly proving the calculus. The "Abstracting simply-typed calculus" section of the chapter discusses the methodology used to apply type to lambda calculus and makes it easier to understand the differences between untyped and simply-typed versions.

Using application

The act of applying one thing to another seems simple enough. When you *apply* peanut butter to toast, you get a peanut butter sandwich. Application in lambda calculus is almost the same thing. If M and N are lambda terms, the combination MN is also a lambda term. In this case, M generally refers to a function and N generally refers to an input to that function, so you often see these terms written as (M)N. The input, N, is applied to the function, M. Because the purpose of the parentheses is to define how to apply terms, it's correct to refer to the pair of parentheses as the *apply operator*.

Understanding that application infers nesting is essential. In addition, because lambda calculus uses only functions, inputs are functions. Consequently, saying $M_2(M_1N)$ would be the same as saying that the function M_1 is applied as input to M_2 and that N is applied as input to M_1.

TECHNICAL
STUFF

In some cases, you see lambda calculus written without the parentheses. For example, you might see EFG as three lambda terms. However, lambda calculus is left associated by default, which means that when you see EFG, what the statement is really telling you is that E is applied to F and F is applied to G, or ((E)F) G. Using the parentheses tends to avoid confusion. Also, be aware that the associative math rule doesn't apply in this case: ((E)F)G is not equivalent to E(F(G)).

To understand the idea of application better, consider the following pseudocode:

```
inc(x) = x + 1
```

All this code means is that to increment x, you add 1 to its value. The lambda calculus form of the same pseudocode written as an anonymous function looks like this:

```
(x) -> x + 1
```

You read this statement as saying that the variable x is mapped to x + 1. However, say that you have a function that requires two inputs, like this:

```
square_sum(x, y) = (x² + y²)
```

The lambda calculus form of the same function written in anonymous form looks like this:

```
(x, y) -> x² + y²
```

This statement is read as saying that the tuple (x, y) is mapped to $x^2 + y^2$. However, as previously mentioned, lambda calculus allows functions to have just one input, and this one has two. To properly apply the functions and inputs, the code would actually need to look like this:

```
x -> (y -> x² + y²)
```

At this point, x and y are mapped separately. The transitioning of the code so that each function has only one argument is called *currying*. This transition isn't precisely how you see lambda calculus written, but it does help explain the underlying mechanisms that you see explained later in the chapter.

Using abstraction

The term *abstraction* derives from the creation of general rules and concepts based on the use and classification of specific examples. The creation of general rules tends to simplify a problem. For example, you know that a computer stores data in memory, but you don't necessarily understand the underlying hardware processes that allow the management of data to take place. The abstraction provided by data storage rules hides the complexity of viewing this process each time it occurs. The following sections describe how abstraction works for both untyped and typed lambda calculus.

Abstracting untyped lambda calculus

In lambda calculus, when E is a lambda term and *x* is a variable, λx.E is a lambda term. An abstraction is a definition of a function, but doesn't invoke the function.

To invoke the function, you must apply it as described in the "Using application" section of the chapter. Consider the following function definition:

```
f(x) = x + 1
```

The lambda abstraction for this function is

```
λx.x + 1
```

Remember that lambda calculus has no concept of a variable declaration. Consequently, when abstracting a function such as

```
f(x) = x² + y²
```

to read

```
λx.x² + y²
```

the variable y is considered a function that isn't yet defined, not a variable declaration. To complete the abstraction, you would create the following:

```
λx.(λy.x² + y²)
```

Abstracting simply-typed calculus

The abstraction process for simply-typed lambda calculus follows the same pattern as described for untyped lambda calculus in the previous section, except that you now need to add type. In this case, the term *type* doesn't refer to string, integer, or Boolean — the types used by other programming paradigms. Rather, *type* refers to the mathematical definition of the function's *domain* (the set of outputs that the function will provide based on its defined argument values) and *range* (the codomain or image of the function), which is represented by A -> B. All that this talk about type really means is that the function can now accept only inputs that provide the correct arguments, and it can provide outputs of only certain arguments as well.

Alonzo Church originally introduced the concept of simply-typed calculus as a simplification of typed calculus to avoid the paradoxical uses of untyped lambda calculus (the "Considering the need for typing" section, later in this chapter, provides details on how this whole process works). A number of lambda calculus extensions (not discussed in this book) also rely on simple typing including: products, coproducts, natural numbers (System T), and some types of recursion (such as Programming Computable Functions, or PCF).

REMEMBER

The important issue for this chapter is how to represent a typed form of lambda calculus statement. For this task, you use the colon (:) to display the expression or variable on the left and the type on the right. For example, referring to the increment abstraction shown in the previous section, you include the type, as shown here:

```
λx:ν.x + 1
```

In this case, the parameter x has a type of ν (nu), which represents natural numbers. This representation doesn't tell the output type of the function, but because + 1 would result in a natural number output as well, it's easy to make the required assumption. This is the Church style of notation. However, in many cases you need to define the type of the function as a whole, which requires the Curry-style notation. Here is the alternative method:

```
(λx.x + 1):ν -> ν
```

Moving the type definition outside means that the example now defines type for the function as a whole, rather than for x. You infer that x is of type ν because the function parameters require it. When working with multiparameter inputs, you must curry the function as shown before. In this case, to assign natural numbers as the type for the sum square function, you might show it like this:

```
λx:ν.(λy:ν.x² + y²)
```

Note the placement of the type information after each parameter. You can also define the function as a whole, like this:

```
(λx.(λy.x² + y²)):ν -> ν -> ν
```

TIP

Each parameter appears separately, followed by the output type. A great deal more exists to discover about typing, but this discussion gives you what you need to get started without adding complexity. The article at http://www.goodmath.org/blog/2014/08/21/types-and-lambda-calculus/ provides additional insights that you may find helpful.

TECHNICAL STUFF

When working with particular languages, you may see the type indicated directly, rather than indirectly using Greek letters. For example, when working with a language that supports the int data type, you may see int used directly, rather than the less direct form of ν that's shown in the previous examples. For example, the following code shows an int alternative to the λx:ν.x + 1 code shown earlier in this section:

```
λx:int.x + 1
```

Performing Reduction Operations

Reduction is the act of expressing a lambda function in its purest, simplest form and ensuring that no ambiguity exists in its meaning. You use one of three kinds of reduction (also called conversion in some cases for the sake of clarity) to perform various tasks in lambda calculus:

» α (alpha)

» β (beta)

» η (eta)

How an application employs these three kinds of reduction is what defines the lambda expression. For example, if you can convert two lambda functions into the same expression, you can consider them β-equivalent. Many texts refer to the process of performing these three kinds of reduction as a whole as λ-reduction. The following sections discuss the reduction operations in detail.

Considering α-conversion

When performing tasks in lambda calculus, you often need to rename variables so that the meaning of the various functions is clear, especially when combining functions. The act of renaming variables is called α-conversion. Two functions are α-equivalent when they have the same result. For example, the following two functions are alpha-equivalent:

```
λx.x + 1
λa.a + 1
```

Clearly, both functions produce the same output, even though one function relies on the letter x, while the second relies on the letter a. However, the alpha-conversion process isn't always straightforward. For example, the following two functions aren't α-equivalent; rather, they're two distinct functions:

```
λx.(λy.x² + y²)
λx.(λx.x² + x²)
```

TIP

Renaming y to x won't work because x is already a captured variable. However, you can rename y to any other variable name desired. In fact, functional language compilers often perform alpha-conversion even when it's not strictly necessary to ensure variable uniqueness throughout an application. Consequently, when viewing your code, you need to consider the effects of alpha-conversion performed solely to ensure variable uniqueness.

Considering β-reduction

The concept of β-reduction is important because it helps simplify lambda functions, sometimes with the help of α-conversion or η-conversion (sometimes called reductions in some texts, but the use of the term *conversion* is clearer). The essential idea is easy—to replace the variables in the body of a function with a particular argument. Making the replacement enables you to solve the lambda function for a particular argument, rather than make a general statement that could apply to any set of arguments.

TIP

The following sections help you understand how beta-reduction works. The tutorial at `http://www.nyu.edu/projects/barker/Lambda/` provides JavaScript aids that can help you to better understand the discussion if you want to feed the examples into the appropriate fields.

Defining bound and unbound variables

When viewing the function, $\lambda x.x + 1$, the λ part of that function binds the variable x to the succeeding expression, $x + 1$. You may also see two other binding operators used in some function declarations: backwards E ($\exists$), also called the existential quantifier used in set theory, or upside-down A ($\forall$), which means *for all*. The examples in this book don't use any of the alternatives, and you don't need to worry about them for now, but you can find a relatively complete set of math symbols, with their explanations, at `https://en.wikipedia.org/wiki/List_of_mathematical_symbols`.

You sometimes find expressions containing unbound or free variables. A free variable is one that appears without the λ part of the function. For example, in this expression, x is bound, while y remains free.

```
λx.x² + y²
```

Unbound variables always remain after any sort of reduction as an unsolved part of the function. It doesn't mean that you won't solve that part of the function — simply that you won't solve it now.

Understanding the basic principle

The previous section discusses bound and unbound variables. This section moves on to the next step: replacing the bound variables with arguments. By performing this step, you move the lambda function from general to specific use. The argument always appears on the right of the function statement. Consequently, the following lambda expression says to apply every argument z to every occurrence of the variable x.

```
((λx.x + 1)z)
```

Notice how the entire function appears in parentheses to separate it from the argument that you want to apply to the function. The result of this reduction appears like this:

```
(z + 1)[x := z]
```

The reduction now shows that the argument z has a value of 1 added to it. The reduction appears in square brackets after the reduced expression. However, the reduction isn't part of the expression; it's merely there for documentation. The whole expression is simply (z + 1).

When performing this task, you need not always use a letter to designate a value. You can use a number or other value instead. In addition, the reduction process can occur in steps. For example, the following reductions require three steps:

```
(((λx.(λy.x² + y²))2)4)
(((λx.(λy1.x² + y1²))2)4)
((λy1.2² + y1²)4)[x := 2]
(2² + 4²)[y1 := 4]
2² + 4²
20
```

This process follows these steps:

1. Use alpha-conversion to rename the y variable to y1 to avoid potential confusion.

2. Replace all occurrences of the x variable with the value 2.

3. Replace all occurrences of the y1 variable with the value 4.

4. Remove the unneeded parentheses.

5. Solve the problem.

Considering the need for typing

The previous section makes beta-reduction look relatively straightforward, even for complex lambda functions. In addition, the previous sections rely on untyped variables. However, untyped variables can cause problems. For example, consider the following beta-reduction:

```
(λx.xx)(λx.xx)
```

Previous examples don't consider two features of this example. First, they didn't consider the potential for using two of the same variable in the expression, xx in this case. Second, they didn't use a function as the argument for the sake of simplicity. To perform the beta-reduction, you must replace each x in the first function with the function that appears as an argument, which results in producing the same code as output. The beta-reduction gets stuck in an endless loop. Now consider this example:

```
L = (λx.xxy)(λx.xxy)
```

The output of this example is (λx.**xxy**)(λx.xxy)y, or Ly, which is actually bigger than before, not reduced. Applying beta-reduction again makes the problem larger still: Lyy. The problem is the fact that the variables have no type, so they accept any input. Adding typing solves this problem by disallowing certain kinds of input. For example, you can't apply a function to this form of the first example:

```
(λx:v.xx)
```

The only argument that will work is a number in this case. Consequently, the function will beta-reduce.

Considering η-conversion

The full implementation of lambda calculus provides a guarantee that the reduction of (λx.Px), in which no argument is applied to x and P doesn't contain x as an unbound (free) variable, results in P. This is the definition of the η-conversion. It anticipates the need for a beta-reduction in the future and makes the overall lambda function simpler before the beta-reduction is needed. The discussion at https://math.stackexchange.com/questions/65622/whats-the-point-of-eta-conversion-in-lambda-calculus provides a fuller look at eta-conversion.

WARNING

The problem with eta-conversion is that few languages actually implement it. Even though eta-conversion should be available, you shouldn't count on its being part of any particular language until you actually test it. For example, the tutorial at http://www.nyu.edu/projects/barker/Lambda/#etareduction shows that eta-conversion isn't available in JavaScript. Consequently, this book doesn't spend a lot of time talking about eta-conversion.

Creating Lambda Functions in Haskell

The "Seeing a Function in Haskell" section of Chapter 4 shows how to create functions in Haskell. For example, if you want to create a curried function to add two numbers together, you might use add x y = x + y. This form of code creates a definite function. However, you can also create anonymous functions in Haskell that rely on lambda calculus to perform a task. The difference is that the function actually is anonymous — has no name — and you assign it to a variable. To see how this process works, open a copy of the Haskell interpreter and type the following code:

```
add = \x -> \y -> x + y
```

Notice how lambda functions rely on the backslash for each variable declaration and the map (->) symbol to show how the variables are mapped to an expression. The form of this code should remind you of what you see in the "Abstracting untyped lambda calculus" section, earlier in this chapter. You now have a lambda function to use in Haskell. To test it, type **add 1 2** and press Enter. The output is 3 as expected.

Obviously, this use of lambda functions isn't all that impressive. You could use the function form without problem. However, lambda functions do come in handy for other uses. For example, you can create specially defined operators. The following code creates a new operator, +=:

```
(+=) = \x -> \y -> x + y
```

To test this code, you type **1+=2** and press Enter. Again, the output is 3, as you might expect. Haskell does allow a shortcut method for defining lambda functions. You can create this same operator using the following code:

```
(+=) = \x y -> x + y
```

Creating Lambda Functions in Python

The "Seeing a Function in Python" section of Chapter 4 shows how to create functions in Python. As with the Haskell function in the previous section of this chapter, you can also create a lambda function version of the add function in Chapter 4. When creating a lambda function in Python, you define the function anonymously and rely on the `lambda` keyword, as shown here:

```
add = lambda x, y: x + y
```

Notice that this particular example assigns the function to a variable. However, you can use a lambda function anywhere that Python expects to see an expression or a function reference. You use this function much as you would any other function. Type **add(1, 2)**, execute the code, and you see 3 as output.

If you want to follow a more precise lambda function formulation, you can create the function like this:

```
add = lambda x: lambda y: x + y
```

In this case, you see how the lambda sequence should work more clearly, but it's extra work. To use this function, you type **add(1)(2)** and execute the code. Python applies the values as you might think, and the code outputs a value of 3.

Python doesn't allow you to create new operators, but you can override existing operators; the article at http://blog.teamtreehouse.com/operator-overloading-python tells you how. However, for this chapter, create a new use for the letter *X* using a lambda function. To begin this process, you must install the Infix module by opening the Anaconda Prompt, typing **pip install infix** at the command prompt, and pressing Enter. After a few moments, pip will tell you that it has installed Infix for you. The following code will let you use the letter *X* to multiply two values:

```
from infix import mul_infix as Infix
X = Infix(lambda x, y: x * y)
5 *X* 6
X(5, 6)
```

The first statement imports `mul_infix` as `Infix`. You have access to a number of infix methods, but this example uses this particular one. The site at https://pypi.org/project/infix/ discusses the other forms of infix at your disposal.

The second statement sets X as the infix function using a lambda expression. The manner in which Infix works allows you to use X as either an operator, as shown by 5 *X* 6 or a regular function, as shown by X(5, 6). When used as an operator, you must surround X with the multiplication operator, *. If you were to use `shif_infix` instead, you would use the shift operators (<< and >>) around the lambda function that you define as the operator.

Chapter **6**

Working with Lists and Strings

C hapter 5 may have given you the idea that the use of lambda calculus in the functional programming paradigm precludes the use of standard program-ming structures in application design. That's not the case, however, and this chapter is here to dispel that myth. In this chapter, you begin with one of the most common and simplest data structures in use today: lists. A *list* is a program-matic representation of the real-world object. Everyone creates lists in real life and for all sorts of reasons. (Just imagine shopping for groceries without a list.) You do the same thing in your applications, even when you're writing code using the functional style. Of course, the functional programming paradigm offers a few surprises, and the chapter discusses them, too.

Sometimes you need to create data structures with greater complexity, which is where the Dict and Set structures come in. Different languages use different terms for these two data structures, but the operation is essentially the same. A Dict offers an ordered list containing name and value pairs. You access the values using the associated name, and the name is what provides the order. A Set offers an unordered collection of elements of the same type with no duplicates. You often use a Set to eliminate duplicate entries from a dataset or to perform mathematical operations such as union and intersection.

The final topic in this chapter involves the use of strings. From a human perspective, strings are an essential means of communicating information. Remember, though, that a computer sees them solely as a string of numbers. Computers work only with numbers, never text, so the representation of a string really is a combination of things that you might not normally think of going together. As with all the other examples in this book, the Haskell and Python string examples use the functional coding paradigm, rather than other paradigms you may have used in the past.

Defining List Uses

After you have used lists, you might be tempted to ask what a list can't do. The list data structure is the most versatile offering for most languages. In most cases, lists are simply a sequence of values that need not be of the same type. You access the elements in a list using an index that begins at 0 for most languages, but could start at 1 for some. The indexing method varies among languages, but accessing specific values using an index is common. Besides storing a sequence of values, you sometimes see lists used in these coding contexts:

>> Stack

>> Queue

>> Deque

>> Sets

Generally, lists offer more manipulation methods than other kinds of data structures simply because the rules for using them are so relaxed. Many of these manipulation methods give lists a bit more structure for use in meeting specialized needs. The "Performing Common List Manipulations" section, later in this chapter, describes these manipulations in detail. Lists are also easy to search and to perform various kinds of analysis. The point is that lists often offer significant flexibility at the cost of absolute reliability and dependability. (You can easily use lists incorrectly, or create scenarios in which lists can actually cause an application to crash, such as when you add an element of the wrong type.)

TIP

Depending on the language you use, lists can provide an impressive array of features and make conversions between types easier. For example, using an iterator in Python lets you perform tasks such as outputting the list as a tuple, processing the content one element at a time, and unpacking the list into separate variables. When working in Haskell, you can create list comprehensions, which are similar in nature to the set comprehensions you work with in math class. The list features you obtain with a particular language depend on the functions the language provides and your own creativity in applying them.

Creating Lists

Before you can use a list, you must create one. Fortunately, most languages make creating lists extremely easy. In some cases, it's a matter of placing a list of values or objects within the correct set of symbols, such as square brackets (which appear to be the most commonly used symbols).

REMEMBER

The most important thing about creating lists is to ensure that you understand how you plan to use the list within the application. Sometimes developers create a freeform list and find out later that controlling the acceptable data types would have been a better idea. Some languages provide methods for ensuring that lists remain pure, but often the ability to control list content is something to add programmatically. The following sections describe how to create lists, first in Haskell and then in Python.

LIST AND ARRAY DIFFERENCE

At first, lists may simply seem to be another kind of array. Many people wonder how lists and arrays differ. After all, from a programming perspective, the two can sound like the same thing. It's true that lists and arrays both store data sequentially, and you can often store any sort of data you want in either structure (although arrays tend to be more restrictive).

The main difference comes in how arrays and lists store the data. An array always stores data in sequential memory locations, which gives an array faster access times in some situations but also slows the creation of arrays. In addition, because an array must appear in sequential memory, updating arrays is often hard, and some languages don't allow you to modify arrays in the same ways that you can lists.

A list stores data using a linked data structure in which a list element consists of the data value and one or two pointers. Lists take more memory because you must now allocate memory for pointers to the next data location (and to the previous location as well in doubly-linked lists, which is the kind used by most languages today). Lists are often faster to create and add data to because of the linking mechanism, but they provide slower read access than arrays.

Using Haskell to create Lists

In Haskell, you can create lists in a number of ways. The easiest method is to define a variable to hold the list and provide the list item within square brackets, as shown here:

```
let a = [1, 2, 3, 4]
```

Notice that the declaration begins with the keyword `let`, followed by a lowercase variable name, which is a in this case. You could also use something more descriptive, such as `myList`. However, if you were to try to use an uppercase beginning letter, you receive an error message like the one shown in Figure 6-1.

FIGURE 6-1:
Variable names must begin with a lowercase letter.

Haskell provides some unique list creation features. For example, you can specify a range of values to put in a list without using any special functions. All you need to do is provide the beginning value, two dots (..), and the ending value, like this:

```
let b = [1..12]
```

You can even use a list comprehension to create a list in Haskell. For example, the following list comprehension builds a list called c based on the doubled content of list a:

```
let c = [x * 2 | x <- a]
```

In this case, Haskell sends the individual values in a to x, doubles the value of x by multiplying by 2, and then places the result in c. List comprehensions give you significant flexibility in creating customized lists. Figure 6-2 shows the output from these two specialized list-creation methods (and many others exist).

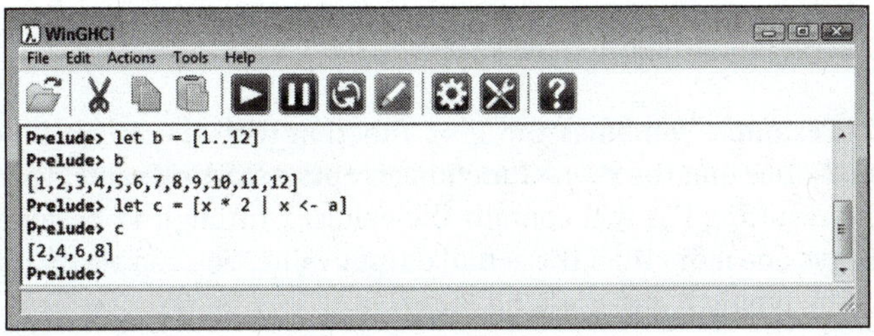

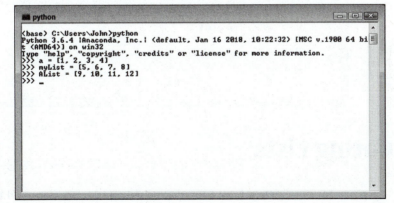

FIGURE 6-2:
Haskell provides specialized list-creation methods.

Using Python to create lists

Creating a list in Python is amazingly similar to creating a list in Haskell. The examples in this chapter are relatively simple, so you can perform them by opening an Anaconda Prompt (a command or terminal window), typing **python** at the command line, and pressing Enter. You use the following code to create a list similar to the one used for the Haskell examples in the previous section:

```
a = [1, 2, 3, 4]
```

In contrast to Haskell variable names, Python variable names can begin with a capital letter. Consequently, the AList example that generates an exception in Haskell, works just fine in Python, as shown in Figure 6-3.

FIGURE 6-3:
Python variable names can begin with an uppercase letter.

You can also create a list in Python based on a range, but the code for doing so is different from that in Haskell. Here is one method for creating a list in Python based on a range:

```
b = list(range(1, 13))
```

This example combines the list function with the range function to create the list. Notice that the range function accepts a starting value, 1, and a stop value, 13. The resulting list will contain the values 1 through 12 because the stop value is always one more than the actual output value. You can verify this output for yourself by typing **b** and pressing Enter.

As does Haskell, Python supports list comprehensions, but again, the code for creating a list in this manner is different. Here's an example of how you could create the list, c, found in the previous example:

```
c = [a * 2 for a in range(1,5)]
```

TIP

This example shows the impure nature of Python because, in contrast to the Haskell example, you rely on a statement rather than lambda calculus to get the job done. As an alternative, you can define the range function stop value by specifying len(a)+1. (The alternative approach makes it easier to create a list based on comprehensions because you don't have to remember the source list length.) When you type **c** and press Enter, the result is the same as before, as shown in Figure 6-4.

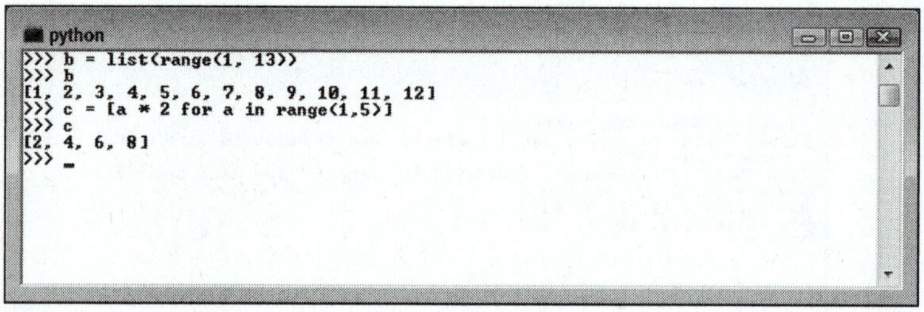

```
python
>>> b = list(range(1, 13))
>>> b
[1, 2, 3, 4, 5, 6, 7, 8, 9, 10, 11, 12]
>>> c = [a * 2 for a in range(1,5)]
>>> c
[2, 4, 6, 8]
>>> _
```

FIGURE 6-4:
Python also allows the use of comprehensions for creating lists.

Evaluating Lists

At some point, you have a list that contains data. The list could be useful at this point, but it isn't actually useful until you evaluate it. *Evaluating* your list means more than simply reading it; it also means ascertaining the value of the list. A list becomes valuable only when the data it contains also become valuable. You can perform this task mathematically, such as determining the minimum, maximum, or mean value of the list, or you can use various forms of analysis to determine how the list affects you or your organization (or possibly a client). Of course, the first step in evaluating a list is to read it.

REMEMBER

This chapter doesn't discuss the process of evaluation fully. In fact, no book can discuss evaluation fully because evaluation means different things to different people. However, the following sections offer enough information to at least start the process, and then you can go on to discover other means of evaluation, including performing analysis (another step in the process) using the techniques described in Part 3 and those provided by other books. The point is that evaluation means to use the data, not to change it in some way. Changes come as part of manipulation later in the chapter.

Using Haskell to evaluate Lists

The previous sections of the chapter showed that you can read a list simply by typing its identifier and pressing Enter. Of course, then you get the entire list. You may decide that you want only part of the list. One way to get just part of the list is to specify an index value, which means using the !! operator in Haskell. To see the first value in a list defined as let a = [1, 2, 3, 4, 5, 6], you type **a !! 0** and press Enter. Indexes begin at 0, so the first value in list a is at index 0, not 1 as you might expect. Haskell actually provides a long list of ways to obtain just parts of lists so that you can see specific elements:

» head a: Shows the value at index 0, which is 1 in this case.

» tail a: Shows the remainder of the list after index 0, which is [2,3,4,5,6] in this case.

» init a: Shows everything except the last element of the list, which is [1,2,3,4,5] in this case.

» last a: Shows just the last element in the list, which is 6 in this case.

» take 3 a: Requires the number of elements you want to see as input and then shows that number from the beginning of the list, which is [1,2,3] in this case.

» drop 3 a: Requires the number of elements you don't want to see as input and then shows the remainder of the list after dropping the required elements, which is [4,5,6] in this case.

USE OF THE GRAVE (BACK QUOTATION MARK OR PRIME) IN HASKELL

Many people will find themselves confused by the use of the accent grave (`` ` ``) or back quotation mark (sometimes called a prime) in Haskell. If you were to type `'div'` (using a single quotation mark instead of the back quotation mark), Haskell would display an error message.

Haskell provides you with a wealth of other ways to slice and dice lists, but it really all comes down to reading the list. The next step is to perform some sort of analysis, which can come in a multitude of ways, but here are some of the simplest functions to consider:

- » `length a`: Returns the number of elements in the list, which is 6 in this case.

- » `null a`: Determines whether the list is empty and returns a Boolean result, which is `False` in this case.

- » `minimum a`: Determines the smallest element of a list and returns it, which is 1 in this case.

- » `maximum a`: Determines the largest element of a list and returns it, which is 6 in this case.

- » `sum a`: Adds the numbers of the list together, which is 21 in this case.

- » `product a`: Multiplies the numbers of the list together, which is 720 in this case.

Haskell does come with an amazing array of statistical functions at `https://hackage.haskell.org/package/statistics`, and you can likely find third-party libraries that offer even more. The "Using Haskell Libraries" section of Chapter 3 tells you how to install and import libraries as needed. However, for something simple, you can also create your own functions. For example, you can use the `sum` and `length` functions to determine the average value in a list, as shown here:

```
avg = \x -> sum(x) `div` length(x)
avg a
```

The output is an integer value of 3 in this case (a sum of 21/6 elements). The lambda function follows the same pattern as that used in Chapter 5. Note that no actual division operator is defined for many operations in Haskell; you use `div` instead. Trying to use something like avg = \x -> sum(x) / length(x) will produce an error. In fact, a number of specialized division-oriented keywords are summarized in the article at https://ebzzry.io/en/division/.

Using Python to evaluate lists

Python provides a lot of different ways to evaluate lists. To start with, you can obtain a particular element using an index enclosed in square brackets. For example, assuming that you have a list defined as a = [1, 2, 3, 4, 5, 6], typing **a[0]** and pressing Enter will produce an output of 1. Unlike in Haskell, you don't have to use odd keywords to obtain various array elements; instead, you use modifications of an index, as shown here:

» a[0]: Obtains the head of the list, which is 1 in this case

» a[1:]: Obtains the tail of the list, which is [2,3,4,5,6] in this case

» a[:-1]: Obtains all but the last element, which is [1,2,3,4,5] in this case

» a[:-1]: Obtains just the last element, which is 6 in this case

» a[:-3]: Performs the same as take 3 a in Haskell

» a[-3:]: Performs the same as drop 3 a in Haskell

As with Haskell, Python probably provides more ways to slice and dice lists than you'll ever need or want. You can also perform similar levels of basic analysis using Python, as shown here:

» len(a): Returns the number of elements in a list.

» not a: Checks for an empty list. This check is different from a is None, which checks for an actual null value — a not being defined.

» min(a): Returns the smallest list element.

» max(a): Returns the largest list element.

» sum(a): Adds the number of the list together.

Interestingly enough, Python has no single method call to obtain the product of a list — that is, all the numbers multiplied together. Python relies heavily on third-party libraries such as NumPy (http://www.numpy.org/) to perform this task.

One of the easiest ways to obtain a product without resorting to a third-party library is shown here:

```
from functools import reduce
reduce(lambda x, y: x * y, a)
```

The `reduce` method found in the `functools` library (see `https://docs.python.org/3/library/functools.html` for details) is incredibly flexible in that you can define almost any operation that works on every element in a list. In this case, the lambda function multiplies the current list element, `y`, by the accumulated value, `x`. If you wanted to encapsulate this technique into a function, you could do so using the following code:

```
prod = lambda z: reduce(lambda x, y: x * y, z)
```

To use `prod` to find the product of list `a`, you would type **prod(a)** and press Enter. No matter how you call it, you get the same output as in Haskell: 720.

Python does provide you with a number of statistical calculations in the `statistics` library (see `https://pythonprogramming.net/statistics-python-3-module-mean-standard-deviation/` for details). However, as in Haskell, you may find that you want to create your own functions to determine things like the average value of the entries in a list. The following code shows the Python version:

```
avg = lambda x: sum(x) // len(x)
avg(a)
```

As before, the output is 3. Note the use of the `//` operator to perform integer division. If you were to use the standard division operator, you would receive a floating-point value as output.

Performing Common List Manipulations

Manipulating a list means modifying it in some way to produce a desired result. A list may contain the data you need, but not the form in which you need it. You may need just part of the list, or perhaps the list is just one component in a larger calculation. Perhaps you don't need a list at all; maybe the calculation requires a tuple instead. The need to manipulate shows that the original list contains something you need, but it's somehow incomplete, inaccurate, or flawed in some other way. The following sections provide an overview of list manipulations that you see enhanced as the book progresses.

Understanding the list manipulation functions

List manipulation means changing the list. However, in the functional programming paradigm, you can't change anything. For all intents and purposes, every variable points to a list that is a constant — one that can't change for any reason whatsoever. So when you work with lists in functional code, you need to consider the performance aspects of such a requirement. Every change you make to any list will require the creation of an entirely new list, and you have to point the variable to the new structure. To the developer, the list may appear to have changed, but underneath, it hasn't — in fact, it can't, or the underlying reason to use the functional programming paradigm fails. With this caveat in mind, here are the common list manipulations you want to consider (these manipulations are in addition to the evaluations described earlier):

» **Concatenation:** Adding two lists together to create a new list with all the elements of both.

» **Repetition:** Creating a specific number of duplicates of a source list.

» **Membership:** Determining whether an element exists within a list and potentially extracting it.

» **Iteration:** Interacting with each element of a list individually.

» **Editing:** Removing specific elements, reversing the list in whole or in part, inserting new elements in a particular location, sorting, or in any other way modifying a part of the list while retaining the remainder.

Using Haskell to manipulate lists

Some of the Haskell list manipulation functionality comes as part of the evaluation process. You simply set the list equal to the result of the evaluation. For example, the following code places a new version of a into b:

```
let a = [1, 2, 3, 4, 5, 6]
let b = take 3 a
```

WARNING

You must always place the result of an evaluation into a new variable. For example, if you were to try using `let a = take 3 a`, as you can with other languages, Haskell would either emit an exception or it would freeze. However, you could later use `a = b` to move the result of the evaluation from `b` to `a`.

Haskell does provide a good supply of standard manipulation functions. For example, `reverse a` would produce `[6,5,4,3,2,1]` as output. You can also split lists using calls such as `splitAt 3 a`, which produces a tuple containing two lists as output: `([1,2,3],[4,5,6])`. To concatenate two lists, you use the concatenation operator: `++`. For example, to concatenate a and b, you use `a ++ b`.

TIP

You should know about some interesting Haskell functions. For example, the `filter` function removes certain elements based on specific criteria, such as all odd numbers. In this case, you use `filter odd a` to produce an output of `[1,3,5]`. The `zip` function is also exceptionally useful. You can use it to combine two lists. Use `zip a ['a', 'b', 'c', 'd', 'e', 'f']` to create a new list of tuples like this: `[(1,'a'),(2,'b'),(3,'c'),(4,'d'),(5,'e'),(6,'f')]`. All these functions appear in the `Data.List` library that you can find discussed at `http://hackage.haskell.org/package/base-4.11.1.0/docs/Data-List.html`.

Using Python to manipulate lists

When working with Python, you have access to a whole array of list manipulation functions. Many of them are dot functions you append to a list. For example, using a list like `a = [1, 2, 3, 4, 5, 6]`, reversing the list would require the `reverse` function like this: `a.reverse()`. However, what you get isn't the output you expected, but a changed version of a. Instead of the original list, a now contains: `[6, 5, 4, 3, 2, 1]`.

Of course, using the dot functions is fine if you want to modify your original list, but in many situations, modifying the original idea is simply a bad idea, so you need another way to accomplish the task. In this case, you can use the following code to reverse a list and place the result in another list without modifying the original:

```
reverse = lambda x: x[::-1]
b = reverse(a)
```

TIP

As with Haskell, Python provides an amazing array of list functions — too many to cover in this chapter (but you do see more as the book progresses). One of the best places to find a comprehensive list of Python list functions is at `https://likegeeks.com/python-list-functions/`.

Understanding the Dictionary and Set Alternatives

This chapter doesn't cover dictionaries and sets in any detail. You use these two structures in detail in Part 3 of the book. However, note that both dictionaries and sets are alternatives to lists and enforce certain rules that make working with data easier because they enforce greater structure and specialization. As mentioned in the chapter introduction, a dictionary uses name/value pairs to make accessing data easier and to provide uniqueness. A set also enforces uniqueness, but without the use of the keys offered by the name part of the name/value pair. You often use dictionaries to store complex datasets and sets to perform specialized math-related tasks.

Using dictionaries

Both Haskell and Python support dictionaries. However, when working with Haskell, you use the HashMap (or a Map). In both cases, you provide name value pairs, as shown here for Python:

```
myDict = {"First": 1, "Second": 2, "Third": 3}
```

The first value, the name, is also a key. The keys are separated from the values by a colon; individual entries are separated by commas. You can access any value in the dictionary using the key, such as print(myDict["First"]). The Haskell version of dictionaries looks like this:

```
import qualified Data.Map as M
let myDict = M.fromList[("First", 1), ("Second", 2),
    ("Third", 3)]

import qualified Data.HashMap.Strict as HM
let myDict2 = HM.fromList[("First", 1), ("Second", 2),
    ("Third", 3)]
```

The Map and HashMap objects are different; you can't interchange them. The two structures are implemented differently internally, and you may find performance differences using one over the other. In creation and use, the two are hard to tell apart. To access a particular member of the dictionary, you use M.lookup "First" myDict for the first and HM.lookup "First" myDict2 for the second. In both cases, the output is Just 1, which indicates that there is only one match and its value is 1. (The discussion at https://stackoverflow.com/questions/7894867/performant-haskell-hashed-structure provides some additional details on how the data structures differ.)

Using sets

Sets in Python are either mutable (the set object) or immutable (the frozenset object). The immutability of the frozenset allows you to use it as a subset within another set or make it hashable for use in a dictionary. (The set object doesn't offer these features.) There are other kinds of sets, too, but for now, the focus is on immutable sets for functional programming uses. Consequently, you see the frozenset used in this book, but be aware that other set types exist that may work better for your particular application. The following code creates a frozenset:

```
myFSet = frozenset([1, 2, 3, 4, 5, 6])
```

You use the frozenset to perform math operations or to act as a list of items. For example, you could create a set consisting of the days of the week. You can't locate individual values in a frozenset but rather must interact with the object as a whole. However, the object is iterable, so the following code tells you whether myFSet contains the value 1:

```
for entry in myFSet:
    if entry == 1:
        print(True)
```

Haskell sets follow a pattern similar to that used for dictionaries. As with all other Haskell objects, sets are immutable, so you don't need to make the same choices as you do when working with Python. The following code shows how to create a set:

```
import Data.Set as Set
let mySet = Set.fromList[1, 2, 3, 4, 5, 6]
```

Oddly enough, the Haskell set is a lot easier to use than the Python set. For example, if you want to know whether mySet contains the value 1, you simply make the following call:

```
Set.member 1 mySet
```

Considering the Use of Strings

Strings convey thoughts in human terms. Humans don't typically speak numbers or math; they use strings of words made up of individual letters to convey thoughts and feelings. Unfortunately, computers don't know what a letter is, much less strings of letters used to create words or groups of words used to create sentences. None of it makes sense to computers. So, as foreign as numbers and math might be to most humans, strings are just as foreign to the computer (if not more so). The following sections provide an overview of the use of strings within the functional programming paradigm.

Understanding the uses for strings

Humans see several kinds of objects as strings, but computer languages usually treat them as separate entities. Two of them are important for programming tasks in this book: characters and strings. A *character* is a single element from a character set, such as the letter A. Character sets can contain nonletter components, such as the carriage return control character. Extended character sets can provide access to letters used in languages other than English. However, no matter how someone structures a character set, a character is always a single entity within that character set. Depending on how the originator structures the character set, an individual character can consume 7, 8, 16, or even 32-bits.

A *string* is a sequential grouping of zero or more characters from a character set. When a string contains zero elements, it appears as an empty string. Most strings contain at least one character, however. The representation of a character in memory is relatively standard across languages; it consumes just one memory location for the specific size of that character. Strings, however, appear in various forms depending on the language. So computer languages treat strings differently from characters because of how each of them uses memory.

Strings don't just see use as user output in applications. Yes, you do use strings to communicate with the user, but you can also use strings for other purposes such as labeling numeric data within a dataset. Strings are also central to certain data formats, such as XML. In addition, strings appear as a kind of data. For example, HTML relies on the agent string to identify the characteristics of the client system. Consequently, even if your application doesn't ever interact with the user, you're likely to use strings in some capacity.

Performing string-related tasks in Haskell

A string is actually a list of characters in Haskell. To see this for yourself, create a new string by typing **let myString = "Hello There!"** and pressing Enter. On the next line, type **:t myString** and press Enter. The output will tell you that myString is of type [Char], a character list.

As you might expect from a purely functional language, Haskell strings are also immutable. When you assign a new string to a Haskell string variable, what you really do is create a new string and point the variable to it. Strings in Haskell are the equivalents of constants in other languages.

Haskell does provide a few string-specific libraries, such as Data.String, where you find functions such as lines (which breaks a string into individual strings in a list between new line characters, \n) and words (which breaks strings into a list of individual words). You can see the results of these functions in Figure 6-5.

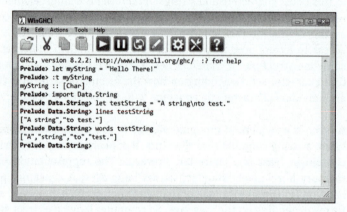

FIGURE 6-5:
Haskell offers at least a few string-related libraries.

REMEMBER

Later chapters spend more time dealing with Haskell strings, but string management is acknowledged as one of the major shortfalls of this particular language. The article at https://mmhaskell.com/blog/2017/5/15/untangling-haskells-strings provides a succinct discussion of some of the issues and demonstrates some string-management techniques. The one thing to get out of this article is that you actually have five different types to deal with if you want to fully implement strings in Haskell.

Performing string-related tasks in Python

Python, as an impure language, also comes with a full list of string functions — too many to go into in this chapter. Creating a string is exceptionally easy: You just type **myString = "Hello There!"** and press Enter. Strings are first-class citizens in Python, and you have access to all the usual manipulation features found in other languages, including special formatting and escape characters. (The tutorial at `https://www.tutorialspoint.com/python/python_strings.htm` doesn't even begin to show you everything, but it's a good start.)

REMEMBER

An important issue for Python developers is that strings are immutable. Of course, that leads to all sorts of questions relating to how someone can seemingly change the value of a string in a variable. However, what really happens is that when you change the contents of a variable, Python actually creates a new string and points the variable to that string rather than the existing string.

TIP

One of the more interesting aspects of Python is that you can also treat strings sort of like lists. The "Using Python to evaluate lists" section talks about how to evaluate lists, and many of the same features work with strings. You have access to all the indexing features to start with, but you can also do things like use `min(myString)`, which returns the space, or `max(myString)`, which returns r, to process your strings. Obviously, you can't use `sum(myString)` because there is nothing to sum. With Python, if you're not quite sure whether something will work on a string, give it a try.

3
Making Functional Programming Practical

Chapter **7**

Performing Pattern Matching

atterns are a set of qualities, properties, or tendencies that form a character-istic or consistent arrangement — a repetitive model. Humans are good at seeing strong patterns everywhere and in everything. In fact, we purposely place patterns in everyday things, such as wallpaper or fabric. However, computers are better than humans are at seeing weak or extremely complex patterns because computers have the memory capacity and processing speed to do so. The capabil-ity to see a pattern is *pattern matching,* which is the overall topic for this chapter. Pattern matching is an essential component in the usefulness of computer systems and has been from the outset, so this chapter is hardly about something radical or new. Even so, understanding how computers find patterns is incredibly important in defining how this seemingly old technology plays such an important part in new applications such as AI, machine learning, deep learning, and data analysis of all sorts.

The most useful patterns are those that we can share with others. To share a pattern with someone else, you must create a language to define it — an *expression.* This chapter also discusses regular expressions, a particular kind of pattern lan-guage, and their use in performing tasks such as data analysis. The creation of a regular expression helps you describe to an application what sort of pattern it should find, and then the computer, with its faster processing power, can locate the precise data you need in a minimum amount of time. This basic information helps you understand more complex pattern matching of the sort that occurs within the realms of AI and advanced data analysis.

Of course, working with patterns using pattern matching through expressions of various sorts works a little differently in the functional programming paradigm. The final sections of this chapter look at how to perform pattern matching using the two languages for this book: Haskell and Python. These examples aren't earth shattering, but they do give you an idea of just how pattern matching works within functional programs so that you can apply pattern matching to other uses.

Looking for Patterns in Data

When you look at the world around you, you see patterns of all sorts. The same holds true for data that you work with, even if you aren't fully aware of seeing the pattern at all. For example, telephone numbers and social security numbers are examples of data that follows one sort of pattern, that of a positional pattern. A telephone number consists of an area code of three digits, an exchange of three digits (even though the exchange number is no longer held by a specific exchange), and an actual number within that exchange of four digits. The positions of these three entities is important to the formation of the telephone number, so you often see a telephone number pattern expressed as (999) 999-9999 (or some variant), where the value 9 is representative of a number. The other characters provide separation between the pattern elements to help humans see the pattern.

Other sorts of patterns exist in data, even if you don't think of them as such. For example, the arrangement of letters from A to Z is a pattern. This may not seem like a revelation, but the use of this particular pattern occurs almost constantly in applications when the application presents data in ordered form to make it easier for humans to understand and interact with the data. Organizational patterns are essential to the proper functioning of applications today, yet humans take them for granted for the most part.

Another sort of pattern is the progression. One of the easiest and most often applied patterns in this category is the exponential progression expressed as N^x, where a number N is raised to the x power. For example, an exponential progression of 2 starting with 0 and ending with 4 would be: 1, 2, 4, 8, and 16. The language used to express a pattern of this sort is the algorithm, and you often use programming language features, such as recursion, to express it in code.

Some patterns are abstractions of real-world experiences. Consider color, as an example. To express color in terms that a computer can understand requires the use of three or four three-digit variables, where the first three are always some value of red, blue, and green. The fourth entry can be an alpha value, which expresses opacity, or a gamma value, which expresses a correction used to define a particular color with the display capabilities of a device in mind. These abstract patterns help humans model the real world in the computer environment so that still other forms of pattern matching can occur (along with other tasks, such as image augmentation or color correction).

Transitional patterns help humans make sense of other data. For example, referencing all data to a known base value makes it possible to compare data from different sources, collected at different times and in different ways, using the same scale. Knowing how various entities collect the required data provides the means for determining which transition to apply to the data so that it can become useful as part of a data analysis.

Data can even have patterns when missing or damaged. The pattern of unusable data could signal a device malfunction, a lack of understanding of how the data collection process should occur, or even human behavioral tendencies. The point is that patterns occur in all sorts of places and in all sorts of ways, which is why having a computer recognize them can be important. Humans may see only part of the picture, but a properly trained computer can potentially see them all.

REMEMBER

So many kinds of patterns exist that documenting them all fully would easily take an entire book. Just keep in mind that you can train computers to recognize and react to data patterns automatically in such a manner that the data becomes useful to humans in various endeavors. The automation of data patterns is perhaps one of the most useful applications of computer technology today, yet very few people even know that the act is taking place. What they see instead is an organized list of product recommendations on their favorite site or a map containing instructions on how to get from one point to another — both of which require the recognition of various sorts of patterns and the transition of data to meet human needs.

Understanding Regular Expressions

Regular expressions are special strings that describe a data pattern. The use of these special strings is so consistent across programming languages that knowing how to use regular expressions in one language makes it significantly easier to use them in all other languages that support regular expressions. As with all reasonably flexible and feature-complete syntaxes, regular expressions can become quite complex, which is why you'll likely spend more than a little time working out the precise manner by which to represent a particular pattern to use in pattern matching.

REMEMBER

You use regular expressions to refer to the technique of performing pattern matching using specially formatted strings in applications. However, the actual code class used to perform the technique appears as Regex, regex, or even RegEx, depending on the language you use. Some languages use a different term entirely, but they're in the minority. Consequently, when referring to the code class rather than the technique, use Regex (or one of its other capitalizations).

TIP

The following sections constitute a brief overview of regular expressions. You can find the more detailed Haskell documentation at `https://wiki.haskell.org/Regular_expressions` and the corresponding Python documentation at `https://docs.python.org/3.6/library/re.html`. These sources of additional help can become quite dense and hard to follow, though, so you might also want to review the tutorial at `https://www.regular-expressions.info/` for further insights.

Defining special characters using escapes

Character escapes usually define a special character of some sort, very often a control character. You escape a character using the backslash (\), which means that if you want to search for a backslash, you must use two backslashes in a row (\\). The character in question follows the escape. Consequently, \b signals that you want to look for a backspace character. Programming languages standardize these characters in several ways:

>> **Control character:** Provides access to control characters such as tab (\t), newline (\n), and carriage return (\r). Note that the \n character (which has a value of \u000D) is different from the \r character (which has a value of \u000A).

>> **Numeric character:** Defines a character based on numeric value. The common types include octal (\nnn), hexadecimal (\xnn), and Unicode (\unnnn). In each case, you replace the *n* with the numeric value of the character, such as \u0041 for a capital letter *A* in Unicode. Note that you must supply the correct number of digits and use 0s to fill out the code.

» **Escaped special character:** Specifies that the regular expression compiler should view a special character, such as (or [, as a literal character rather than as a special character. For example, \(would specify an opening parenthesis rather than the start of a subexpression.

Defining wildcard characters

A wildcard character can define a kind of character, but never a specific character. You use wildcard characters to specify any digit or any character at all. The following list tells you about the common wildcard characters. Your language may not support all these characters, or it may define characters in addition to those listed. Here's what the following characters match with:

» **.:** Any character (with the possible exception of the newline character or other control characters).

» **\w:** Any word character

» **\W:** Any nonword character

» **\s:** Any whitespace character

» **\S:** Any non-whitespace character

» **\d:** Any decimal digit

» **\D:** Any nondecimal digit

Working with anchors

Anchors define how to interact with a regular expression. For example, you may want to work only with the start or end of the target data. Each programming language appears to implement some special conditions with regard to anchors, but they all adhere to the basic syntax (when the language supports the anchor). The following list defines the commonly used anchors:

» **^:** Looks at the start of the string.

» **$:** Looks at the end of the string.

» ***:** Matches zero or more occurrences of the specified character.

» **+:** Matches one or more occurrences of the specified character. The character must appear at least once.

» **?:** Matches zero or one occurrences of the specified character.

>> **{*m*}:** Specifies *m* number of the preceding characters required for a match.

>> **{*m,n*}:** Specifies the range from *m* to *n* number of the preceding characters required for a match.

>> **expression|expression:** Performs or searches where the regular expression compiler will locate either one expression or the other expression and count it as a match.

REMEMBER

You may find figuring out some of these anchors difficult. The idea of matching means to define a particular condition that meets a demand. For example, consider this pattern: h?t, which would match *hit* and *hot*, but not *hoot* or *heat*, because the ? anchor matches just one character. If you instead wanted to match *hoot* and *heat* as well, you'd use h*t, because the * anchor can match multiple characters. Using the right anchor is essential to obtaining a desired result.

Delineating subexpressions using grouping constructs

A grouping construct tells the regular expression compiler to treat a series of characters as a group. For example, the grouping construct [a–z] tells the regular expression compiler to look for all lowercase characters between *a* and *z*. However, the grouping construct [az] (without the dash between *a* and *z*) tells the regular expression compiler to look for just the letters *a* and *z*, but nothing in between, and the grouping construct [^a–z] tells the regular expression compiler to look for everything but the lowercase letters *a* through *z*. The following list describes the commonly used grouping constructs. The italicized letters and words in this list are placeholders.

>> **[*x*]:** Look for a single character from the characters specified by *x*.

>> **[*x-y*]:** Search for a single character from the range of characters specified by *x* and *y*.

>> **[^expression]:** Locate any single character not found in the *character expression*.

>> **(expression):** Define a regular *expression* group. For example, ab{3} would match the letter *a* and then three copies of the letter *b*, that is, abbb. However, (ab){3} would match three copies of the expression ab: ababab.

Using Pattern Matching in Analysis

Pattern matching in computers is as old as the computers themselves. In looking at various sources, you can find different starting points for pattern matching, such as editors. However, the fact is that you can't really do much with a computer system without having some sort of pattern matching occur. For example, the mere act of stopping certain kinds of loops requires that a computer match a pattern between the existing state of a variable and the desired state. Likewise, user input requires that the application match the user's input to a set of acceptable inputs.

Developers recognize that function declarations also form a kind of pattern and that in order to call the function successfully, the caller must match the pattern. Sending the wrong number or types of variables as part of the function call causes the call to fail. Data structures also form a kind of pattern because the data must appear in a certain order and be of a specific type.

REMEMBER

Where you choose to set the beginning for pattern matching depends on how you interpret the act. Certainly, pattern matching isn't the same as counting, as in a for loop in an application. However, it could be argued that testing for a condition in a while loop matches the definition of pattern matching to some extent. The reason that many people look at editors as the first use of pattern matching is that editors were the first kinds of applications to use pattern matching to perform a search, such as to locate a name in a document. Searching is most definitely part of the act of analysis because you must find the data before you can do anything with it.

The act of searching is just one aspect, however, of a broader application of pattern matching in analysis. The act of filtering data also requires pattern matching. A search is a singular approach to pattern matching in that the search succeeds the moment that the application locates a match. Filtering is a batch process that accepts all the matches in a document and discards anything that doesn't match, enabling you to see all the matches without doing anything else. Filtering can also vary from searching in that searching generally employs static conditions, while filtering can employ some level of dynamic condition, such as locating the members of a set or finding a value within a given range.

Filtering is the basis for many of the analysis features in declarative languages, such as SQL, when you want to locate all the instances of a particular data structure (a record) in a large data store (the database). The level of filtering in SQL is much more dynamic than in mere filtering because you can now apply conditional sets and limited algorithms to the process of locating particular data elements.

Regular expressions, although not the most advanced of modern pattern-matching techniques, offer a good view of how pattern matching works in modern applications. You can check for ranges and conditional situations, and you can even apply a certain level of dynamic control. Even so, the current master of pattern matching is the algorithm, which can be fully dynamic and incredibly responsive to particular conditions.

Working with Pattern Matching in Haskell

Haskell provides a full array of pattern matching functionality. The example in this section specifically uses the `Text.Regex.Posix` package. You can, in fact, find a number of regular expression implementations discussed at https://wiki.haskell.org/Regular_expressions. However, the implementation that is easiest to use is the `Text.Regex.Posix` package described at http://hackage.haskell.org/package/regex-posix-0.95.1/docs/Text-Regex-Posix.html and supported by the `Text.Regix` package described at http://hackage.haskell.org/package/regex-compat-0.95.1/docs/Text-Regex.html. The following sections detail Haskell pattern matching using two examples.

Performing simple Posix matches

Every language you work with will have quirks when it comes to Regex. Unfortunately, figuring out what those quirks are in Haskell can prove frustrating at times. One of the better resources you can use to determine how to format a Haskell Regex string is at https://www.regular-expressions.info/posix.html. In fact, if you look at the sidebar for this site, you find Regex implementations for a lot of other languages as well. The `Text.Regex.Posix` package adheres to these conventions for the most part. However, when looking at the table used to describe character classes at https://www.regular-expressions.info/posixbrackets.html, you need to know that Haskell doesn't support the shorthand characters, such as \d for digits. You must use [0-9] to represent all digits instead.

To begin working with the Posix form of Regex in Haskell, you first type **import Text.Regex.Posix** and press Enter. You can then create a pattern to use, such as `let vowels = "[aeiou]"`. Try it by typing **"This is a test sentence." =~ vowels ::** **Bool** and pressing Enter. The results appear in Figure 7-1.

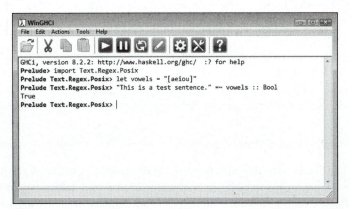

FIGURE 7-1:
This Haskell Regex checks a test sentence for the presence of vowels.

The output shows that the sentence does contain vowels. Notice that the test process uses the =~ (that's a tilde, ~) operator. In addition, you choose a form of output by providing the :: operator followed by a type, which is Bool in this case. The results are interesting if you change types. For example, if you use Int instead, you discover that the test sentence contains seven vowels. Using String tells you that the first instance is the letter *i*.

TIP

Things become even more interesting when you provide tuples as the type. For example, if you use (String,String,String), you see the entire sentence with the part before the match as the first entry, the match itself, and then the part after the match as the last entry, as shown in Figure 7-2. The (String,String, String,[String]) tuple provides the addition of a list of matching groups.

FIGURE 7-2:
Provide a tuple as input to get a fuller view of how the match appears in context.

Another useful tuple is (Int,Int). In this case, you receive the starting zero-based offset of the first match and its length. In this case, you see an output of (2,1) because i is at offset 2 and has a length of 1.

Matching a telephone number with Haskell

The previous section outlined the methods used to create many simple matches in Haskell. However, most matches aren't all that simple. The example in this section shows a common type of match that requires a bit more in the way of expression string structure: a telephone number. Here's the pattern used for this section:

```
let tel = "\\([0-9]{3}\\)[0-9]{3}\\-[0-9]{4}"
```

The `tel` pattern includes several new features. Beginning at the left, you have the `\\` escape that equates to a literally interpreted backslash, followed by a left parenthesis, `(`. You need all three of these characters to define an opening parenthesis in the pattern. The next entry is the numbers 0 through 9 (`[0-9]`) repeated three times (`{3}`). The next three characters define the closing parenthesis for the area code. Defining the exchange, `[0-9]{3}`, comes next. To create a dash between the exchange and the number, you must again use a three-character combination of `\\-`. The number part of the pattern appears as `[0-9]{4}`.

To test this pattern out, type **"My telephone number is: (555)555-1234." =~ tel :: String** and press Enter. (The `My telephone number is:` part of the entry isn't part of the pattern, and you'll get the right output even if you don't include it.) You see just the telephone number as output. Of course, you can use all the type modifiers discussed in the previous section. The problem with using pattern matching is that it can be quite brittle. If you instead type **"My telephone number is: 555-555-1234." =~ tel :: String** and press Enter, you get a blank output, despite the fact that the sentence does include what humans would recognize as a telephone number, as shown in Figure 7-3. The problem is that the pattern doesn't match this form of telephone number.

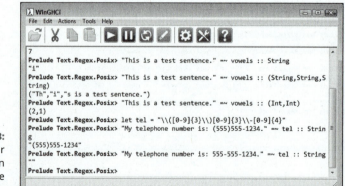

FIGURE 7-3: Regular expression patterns can be brittle.

Working with Pattern Matching in Python

Pattern matching in Python closely matches the functionality found in many other languages. Although Haskell (discussed in the previous section) can seem a little limited, Python more than makes up for it with robust pattern-matching capabilities provided by the regular expression (re) library (https://docs.python.org/3.6/library/re.html). The resource at https://www.regular-expressions.info/python.html provides a good overview of the Python capabilities. The following sections detail Python functionality using a number of examples.

Performing simple Python matches

All the functionality you need for employing Python in basic RegEx tasks appears in the re library. To use this library, you type **import re** and press Enter. As with Haskell, you can create a pattern by typing **vowels = "[aeiou]"** and pressing Enter. Test the result by typing **re.search(vowels, "This is a test sentence.")** and pressing Enter. The only problem is that you get a match object (https://docs.python.org/3.6/library/re.html#match-objects) in return, rather than the actual search value as you might have expected. To overcome this issue, make a call to group, re.search(vowels, "This is a test sentence.").group(). Figure 7-4 shows the difference.

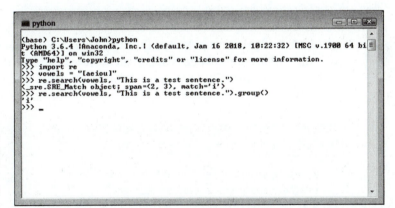

When you look at the Python documentation, you find quite a few functions devoted to working with regular expressions, some of them not entirely clear in their purpose. For example, you have a choice between performing a search or a match. A match works only at the beginning of a string. Consequently, if you type

re.match(vowels, "This is a test sentence.") and press Enter, you see no output at all, which seems impossible given that there should be a match. To understand the difference, type **re.match("a", "abcde")** and press Enter. Now you see a match because the match occurs at the first letter of the target string.

Neither search nor match will locate all occurrences of the pattern in the target string. To locate all the matches, you use findall or finditer instead. To see how this works, type **re.findall(vowels, "This is a test sentence.")** and press Enter. You see the list output shown in Figure 7-5. Because this is a list, you can manipulate it as you would any other list.

```
python
(base) C:\Users\John>python
Python 3.6.4 |Anaconda, Inc.| (default, Jan 16 2018, 10:22:32) [MSC v.1900 64 bi
t (AMD64)] on win32
Type "help", "copyright", "credits" or "license" for more information.
>>> import re
>>> vowels = "[aeiou]"
>>> re.search(vowels, "This is a test sentence.")
<_sre.SRE_Match object; span=(2, 3), match='i'>
>>> re.search(vowels, "This is a test sentence.").group()
'i'
>>> re.match(vowels, "This is a test sentence.")
>>> re.match("a", "abcde")
<_sre.SRE_Match object; span=(0, 1), match='a'>
>>> re.findall(vowels, "This is a test sentence.")
['i', 'i', 'a', 'e', 'e', 'e', 'e']
>>> 
```

FIGURE 7-5:
Use findall to locate all the matches for a pattern in a string.

Look again at Figure 7-4. Notice that the match object contains the entry span=(2, 3). That information is important because it tells you the location of the match in the sentence. You can use this information with the match object start and end functions, as shown here:

```
testSentence = "This is a test sentence."
m = re.search(vowels, testSentence)
while m:
    print(testSentence[m.start():m.end()])
    testSentence = testSentence[m.end():]
    m = re.search(vowels, testSentence)
```

This code keeps performing searches on the remainder of the sentence after each search until it no longer finds a match. Figure 7-6 shows the output from this example. Obviously, using the finditer function would be easier, but this code points out that Python does provide everything needed to create relatively complex pattern-matching code.

FIGURE 7-6:
Python makes relatively complex pattern-matching sequences possible.

Doing more than matching

Python's regular expression library makes it quite easy to perform a wide variety of tasks that don't strictly fall into the category of pattern matching. This chapter discusses only a few of the more interesting capabilities. One of the most commonly used is splitting strings. For example, you might use the following code to split a test string using a number of whitespace characters:

```
testString = "This is\ta test string.\nYippee!"
whiteSpace = "[\s]"
re.split(whiteSpace, testString)
```

The escaped character, \s, stands for all space characters, which includes the set of [\t\n\r\f\v]. The split function can split any content using any of the accepted regular expression characters, so it's an extremely powerful data manipulation function. The output from this example appears in Figure 7-7.

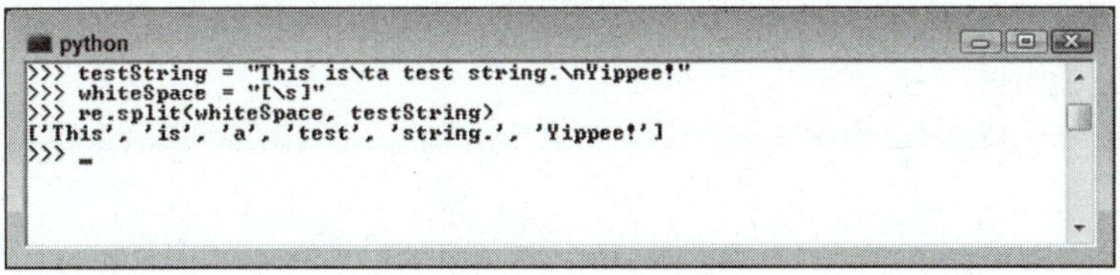

FIGURE 7-7:
The split function provides an extremely powerful method for manipulating data.

```
>>> testString = "This is\ta test string.\nYippee!"
>>> whiteSpace = "[\s]"
>>> re.split(whiteSpace, testString)
['This', 'is', 'a', 'test', 'string.', 'Yippee!']
>>>
```

Performing substitutions using the sub function is another forte of Python. Rather than perform common substitutions one at a time, you can perform them all simultaneously, as long as the replacement value is the same in all cases. Consider the following code:

```
testString = "Stan says hello to Margot from Estoria."
pattern = "Stan|hello|Margot|Estoria"
replace = "Unknown"
re.sub(pattern, replace, testString)
```

The output of this example is 'Unknown says Unknown to Unknown from Unknown.'. You can create a pattern of any complexity and use a single replacement value to represent each match. This is handy when performing certain kinds of data manipulation for tasks such as dataset cleanup prior to analysis.

Matching a telephone number with Python

Whether you create a telephone number matching a regular expression in Python or Haskell, the same basic principles apply. The language-specific implementation details, however, do differ. When creating the pattern in Python, you type something along the lines of **tel = "\(\d{3}\)\d{3}\-\d{4}"** and press Enter. Note that as with Haskell, you must escape the (,), and – characters. However, in contrast to Haskell, you do have access to the shortcuts. The \d represents any digit.

To test this pattern, type **re.search(tel, testString).group()** and press Enter. You see the output shown in Figure 7-8. As with the Haskell example, this pattern is equally brittle, and you would need to amend it to make it work with a variety of telephone-number patterns.

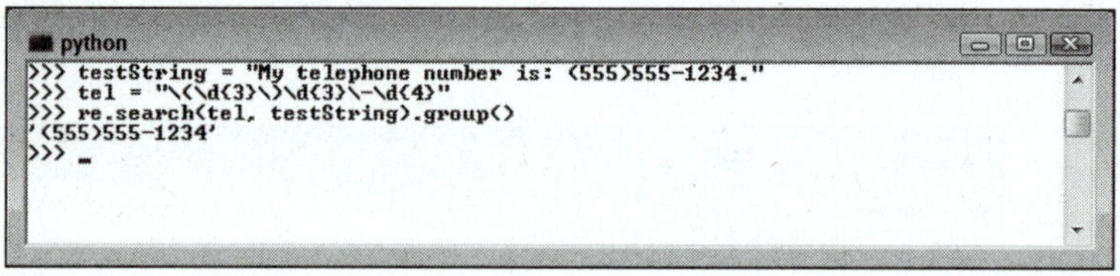

FIGURE 7-8:
Python makes working with patterns a little easier than Haskell does.

```
python
>>> testString = "My telephone number is: (555)555-1234."
>>> tel = "\(\d{3}\)\d{3}\-\d{4}"
>>> re.search(tel, testString).group()
'(555)555-1234'
>>>
```

Chapter **8**

Using Recursive Functions

Some people confuse recursion with a kind of looping. The two are completely different sorts of programming and wouldn't even look the same if you could view them at a low level. In *recursion,* a function calls itself repetitively and keeps track of each call through stack entries, rather than an application state, until the condition used to determine the need to make the function call meets some requirement. At this point, the list of stack entries unwinds with the function passing the results of its part of the calculation to the caller until the stack is empty and the initiating function has the required output of the call. Although this sounds mind-numbingly complex, in practice, recursion is an extremely elegant method of solving certain computing problems and may be the only solution in some situations. This chapter introduces you to the basics of recursion using the two target languages for this book, so don't worry if this initial definition leaves you in doubt of what recursion means.

Of course, you might wonder just how well recursion works on some common tasks, such as iterating a list, dictionary, or set. This chapter discusses all the basic requirements for replacing loops with recursion when interacting with common data structures. It then moves on to more advanced tasks and demonstrates how using recursion can actually prove superior to the loops you used in the past — not to mention that it can be easier to read and more flexible as well. Of course, when using other programming paradigms, you'll likely continue using loops because those languages are designed to work with them.

REMEMBER

One of the most interesting aspects of using first-class functions in the functional programming paradigm is that you can now pass functions rather than variables to enable recursion. This capability makes recursion in the functional programming paradigm significantly more powerful than it is in other paradigms because the function can do more than a variable, and by passing different functions, you can alter how the main recursion sequence works.

People understand loops with considerably greater ease because we naturally use loops in our daily lives. Every day, you perform repetitive tasks a set number of times or until you satisfy a certain condition. You go to the store, know that you need a dozen nice apples, and count them out one at a time as you place them in one of those plastic bags. The lack of recursion in our daily lives is one of the reasons that it's so hard to wrap our brains around recursion, and it's why people make so many common mistakes when using recursion. The end of the chapter discusses a few common programming errors so that you can successfully use recursion to create an amazing application.

Performing Tasks More than Once

One of the main advantages of using a computer is its capability to perform tasks repetitively — often far faster and with greater accuracy than a human can. Even a language that relies on the functional programming paradigms requires some method of performing tasks more than once; otherwise, creating the language wouldn't make sense. Because the conditions under which functional languages repeat tasks differ from those of other languages using other paradigms, thinking about the whole concept of repetition again is worthwhile, even if you've worked with these other languages. The following sections provide you with a brief overview.

Defining the need for repetition

The act of repeating an action seems simple enough to understand. However, repetition in applications occurs more often than you might think. Here are just a few uses of repetition to consider:

>> Performing a task a set number of times

>> Performing a task a variable number of times until a condition is met

>> Performing a task a variable number of times until an event occurs

>> Polling for input

>> Creating a message pump

>> Breaking a large task into smaller pieces and then executing the pieces

>> Obtaining data in chunks from a source other than the application

>> Automating data processing using various data structures as input

In fact, you could easily create an incredibly long list of repeated code elements in most applications. The point of repetition is to keep from writing the same code more than once. Any application that contains repeated code becomes a maintenance nightmare. Each routine must appear only once to make its maintenance easy, which means using repetition to allow execution more than one time.

TIP

Without actual loop statements, the need for repetition becomes significantly clearer because repetition suddenly receives the limelight. Thinking through the process of repeating certain acts without relying on state or mutable variables takes on new significance. In fact, the difficulty of actually accomplishing this task in a straightforward manner forces most language designers to augment even pure languages with some sort of generalized looping mechanism. For example, Haskell provides the forM function, which actually has to do with performing I/O (see the article at http://learnyouahaskell.com/input-and-output for details). The Control.Monad library contains a number of interesting loop-like functions that really aren't loops; they're functions that implement repetition using data structures as input (see https://hackage.haskell.org/package/base-4.2.0.1/docs/Control-Monad.html for details). Here's an example of forM in use:

```
import Control.Monad
values <- forM [1, 2, 3, 4]
    (\a -> do
       putStrLn $ "The value is: " ++ show a )
```

In this case, forM processes a list containing four values, passing it to the lambda function that follows. The lambda function simply outputs the values using putStrLn and show a. Figure 8-1 shows the output from this example. Obviously, impure languages, such as Python, do provide the more traditional methods of repeating actions.

Using recursion instead of looping

The functional programming paradigm doesn't allow the use of loops for two simple reasons. First, a loop requires the maintenance of state, and the functional programming paradigm doesn't allow state. Second, loops generally require mutable variables so that the variable can receive the latest data updates as the loop continues to perform its task. As mentioned previously, you can't use mutable variables in functional programming. These two reasons would seem to sum up the entirety of why to avoid loops, but there is yet another.

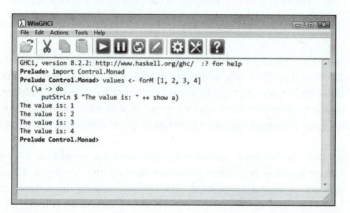

FIGURE 8-1:
Even Haskell provides a looping-type mechanism in the form of functions.

One of the reasons that functional programming is so amazing is that you can use it on multiple processors without concern for the usual issues found with other programming paradigms. Because each function call is guaranteed to produce precisely the same result, every time, given the same inputs, you can execute a function on any processor without regard to the processor use for the previous call. This feature also affects recursion because recursion lacks state.

When a function calls itself, it doesn't matter where the next function call occurs; it can occur on any processor in any location. The lack of state and mutable variables makes recursion the perfect tool for using as many processors as a system has to speed applications as much as possible.

Understanding Recursion

Recursion, in its essence, is a method of performing tasks repetitively, wherein the original function calls itself. Various methods are available for accomplishing this task, as described in the following sections. The important aspect to keep in mind, though, is the repetition. Whether you use a list, dictionary, set, or collection as the mechanism to input data is less important than the concept of a function's calling itself until an event occurs or it fulfills a specific requirement.

Considering basic recursion

This section discusses basic recursion, which is the kind that you normally see demonstrated for most languages. In this case, the doRep function creates a list containing a specific number, n, of a value, x, as shown here for Python:

```
def doRep(x, n):
    y = []
    if n == 0:
        return []
    else:
        y = doRep(x, n - 1)
        y.append(x)
        return y
```

To understand this code, you must think about the process in reverse. Before it does anything else, the code actually calls doRep repeatedly until n == 0. So, when n == 0, the first actual step in the recursion process is to create an empty list, which is what the code does.

At this point, the call returns and the first actual step concludes, even though you have called doRep six times before it gets to this point. The next actual step, when n == 1, is to make y equal to the first actual step, an empty list, and then append x to the empty list by calling y.append(x). At this point, the second actual step concludes by returning [4] to the previous step, which has been waiting in limbo all this time.

The recursion continues to unwind until n == 5, at which point it performs the final append and returns [4, 4, 4, 4, 4] to the caller, as shown in Figure 8-2.

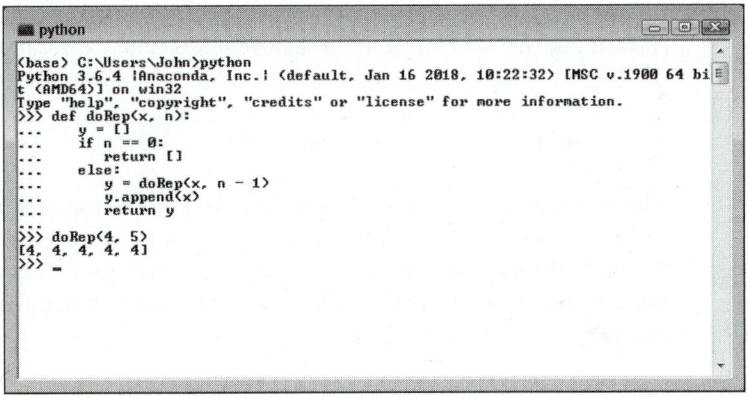

FIGURE 8-2:
Recursion almost seems to work backward when it comes to code execution.

TIP

Sometimes it's incredibly hard to wrap your head around what happens with recursion, so putting a print statement in the right place can help. Here's a modified version of the Python code with the print statement inserted. Note that the print statement goes after the recursive call so that you can see the result of making it. Figure 8-3 shows the flow of calls in this case.

```python
def doRep(x, n):
    y = []
    if n == 0:
        return []
    else:
        y = doRep(x, n - 1)
        print(y)
        y.append(x)
        return y
```

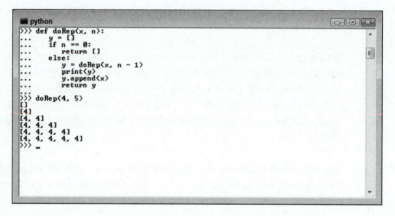

FIGURE 8-3:
Adding a print statement in the right place can make recursion easier to understand.

Performing the same task in Haskell requires about the same code, but with a Haskell twist. Here's the same function written in Haskell:

```
doRep x n | n <= 0 = [] | otherwise = x:doRep x (n–1)
```

The technique is the same as the Python example. The function accepts two inputs, x (the number to append) and n (the number of times to append it). When n is 0, the Haskell code returns an empty list. Otherwise, Haskell appends x to the list and returns the appended list. Figure 8-4 shows the functionality of the Haskell example.

FIGURE 8-4:
Haskell also
makes creating
the doRep
function easy.

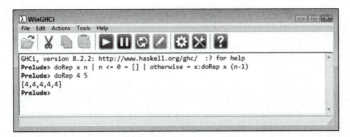

Performing tasks using lists

Lists represent multiple inputs to the same call during the same execution. A list can contain any data type in any order. You use a list when a function requires more than one value to calculate an output. For example, consider the following Python list:

```
myList = [1, 2, 3, 4, 5]
```

If you wanted to use standard recursion to sum the values in the list and provide an output, you could use the following code:

```
def lSum(list):
    if not list:
        return 0
    else:
        return list[0] + lSum(list[1:])
```

The function relies on slicing to remove one value at a time from the list and add it to the sum. The *base case* (principle, simplest, or foundation) is that all the values are gone and now list contains the empty set, ([]), which means that it has a value of 0. To test this example out, you type **lSum(myList)** and press Enter.

Using lambda functions in Python recursion isn't always easy, but this particular example lends itself to using a lambda function quite easily. The advantage is that you can create the entire function in a single line, as shown in the following code (with two lines, but using a line-continuation character):

```
lSum2 = lambda list: 0 if not list \
    else list[0] + lSum2(list[1:])
```

The code works precisely the same as the longer example, relying on slicing to get the job done. You use it in the same way, typing **lSum2(myList)** and pressing Enter. The result is the same, as shown in Figure 8-5.

Upgrading to set and dictionary

Both Haskell and Python support sets, but with Python, you get them as part of the initial environment, and with Haskell, you must load the Data.Set library (see http://hackage.haskell.org/package/containers-0.5.11.0/docs/Data-Set.html to obtain the required support). *Sets* differ from lists in that sets can't contain any duplicates and are generally presented as ordered. In Python, the sets are stored as hashes. In some respects, sets represent a formalization of lists. You can convert lists to sets in either language. Here's the Haskell version:

```
import Data.Set as Set
myList = [1, 4, 8, 2, 3, 3, 5, 1, 6]
mySet = Set.fromList(myList)
mySet
```

which has an output of fromList [1,2,3,4,5,6,8] in this case. Note that even though the input list is unordered, mySet is ordered when displayed. Python relies on the set function to perform the conversion, as shown here:

```
myList = [1, 4, 8, 2, 3, 3, 5, 1, 6]
mySet = set(myList)
mySet
```

The output in this case is {1, 2, 3, 4, 5, 6, 8}. Notice that sets in Python use a different enclosing character, the curly brace. The unique and ordered nature of sets makes them easier to use in certain kinds of recursive routines, such as finding a unique value.

You can find a number of discussions about Haskell sets online, and some people are even unsure as to whether the language implements them as shown at `https://stackoverflow.com/questions/7556573/why-is-there-no-built-in-set-data-type-in-haskell`. Many Haskell practitioners prefer to use lists for everything and then rely on an approach called *list comprehensions* to achieve an effect similar to using sets, as described at `http://learnyouahaskell.com/starting-out#im-a-list-comprehension`. The point is that if you want to use sets in Haskell, they are available.

A *dictionary* takes the exclusivity of sets one step further by creating key/value pairs, in which the key is unique but the value need not be. Using keys makes searches faster because the keys are usually short and you need only to look at the keys to find a particular value. Both Haskell and Python place the key first, followed by the value. However, the methods used to create a dictionary differ. Here's the Haskell version:

```
let myDic = [("a", 1), ("b", 2), ("c", 3), ("d", 4)]
```

Note that Haskell actually uses a list of tuples. In addition, many Haskell practitioners call this an *association list*, rather than a dictionary, even though the concept is the same no matter what you call it. Here's the Python form:

```
myDic = {"a": 1, "b": 2, "c": 3, "d": 4}
```

Python uses a special form of set to accomplish the same task. In both Python and Haskell, you can access the individual values by using the key. In Haskell, you might use the `lookup` function: `lookup "b" myDic`, to find that b is associated with 2. Python uses a form of index to access individual values, such as `myDic["b"]`, which also accesses the value 2.

You can use recursion with both sets and dictionaries in the same manner as you do lists. However, recursion really begins to shine when it comes to complex data structures. Consider this Python nested dictionary:

```
myDic = {"A":{"A": 1, "B":{"B": 2, "C":{"C": 3}}}, "D": 4}
```

In this case, you have a dictionary nested within other dictionaries down to four levels, creating a complex dataset. In addition, the nested dictionary contains the same "A" key value as the first level dictionary (which is allowed), the same "B" key value as the second level, and the "C" key on the third level. You might need to look for the repetitious keys, and recursion is a great way to do that, as shown here:

```python
def findKey(obj, key):
    for k, v in obj.items():
        if isinstance(v, dict):
            findKey(v, key)
        else:
            if key in obj:
                print(obj[key])
```

This code looks at all the entries by using a for loop. Notice that the loop unpacks the entries into key, k, and value, v. When the value is another dictionary, the code recursively calls findKey with the value and the key as input. Otherwise, if the instance isn't a dictionary, the code checks to verify that the key appears in the input object and prints just the value of that object. In this case, an object can be a single entry or a sub-dictionary. Figure 8-6 shows this example in use.

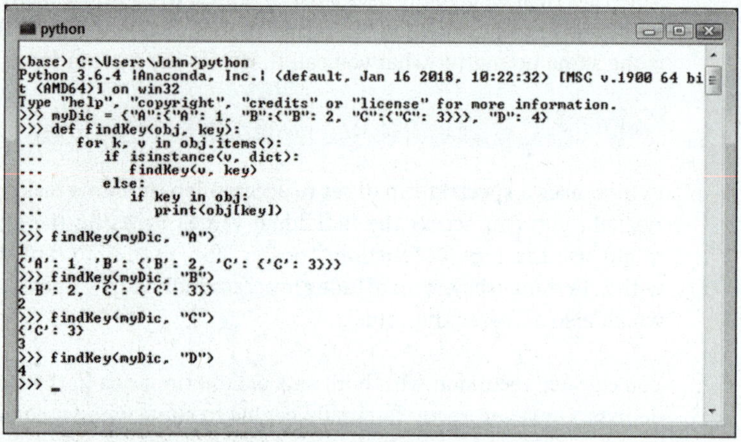

FIGURE 8-6: Dictionaries can provide complex datasets you can parse recursively.

Considering the use of collections

Depending on which language you choose, you might have access to other kinds of collections. Most languages support lists, sets, and dictionaries as a minimum, but you can see some alternatives for Haskell at http://hackage.haskell.org/package/collections-api-1.0.0.0/docs/Data-Collections.html and Python at https://docs.python.org/3/library/collections.html. All these collections have one thing in common: You can use recursion to process their content. Often, the issue isn't one of making recursion work, but of simply finding the right technique for accessing the individual data elements.

Using Recursion on Lists

The previous sections of the chapter prepare you to work with various kinds of data structures by using recursion. You commonly see lists used whenever possible because they're simple and can hold any sort of data. Unfortunately, lists can also be quite hard to work with because they can hold any sort of data (most analysis requires working with a single kind of data) and the data can contain duplicates. In addition, you might find that the user doesn't provide the right data at all. The following sections look at a simple case of looking for the next and previous values in the following list of numbers:

```
myList = [1,2,3,4,5,6]
```

Working with Haskell

This example introduces a few new concepts from previous examples because it seeks to provide more complete coverage. The following code shows how to find the next and previous values in a list:

```
findNext :: Int -> [Int] -> Int
findNext _ [] = -1
findNext _[_] = -1
findNext n (x:y:xs)
    | n == x = y
    | otherwise = findNext n (y:xs)

findPrev :: Int -> [Int] -> Int
findPrev _ [] = -1
findPrev _[_] = -1
findPrev n (x:y:xs)
    | n == y = x
    | otherwise = findPrev n (y:xs)
```

In both cases, the code begins with a type signature that defines what is expected regarding inputs and outputs. As you can see, both functions require Int values for input and provide Int values as output. The next two lines provide outputs that handle the cases of empty input (a list with no entries) and singleton input (a list with one entry). There is no next or previous value in either case.

The meat of the two functions begins by breaking the list into parts: the current entry, the next entry, and the rest of the list. So, the processing begins with $x = 1, y = 2$, and $xs = [3,4,5,6]$. The code begins by asking whether n equals x for findNext and whether n equals y for findPrev. When the two are equal, the function returns either the next or previous value, as appropriate. Otherwise, it recursively calls findNext or findPrev with the remainder of the list. Because the list gets one item shorter with each recursion, processing stops with a successful return of the next or previous value, or with –1 if the list is empty. Figure 8-7 shows this example in use. The figure presents both successful and unsuccessful searches.

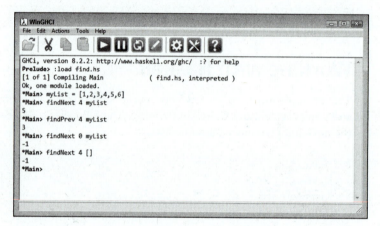

FIGURE 8-7: The findNext and findPrev functions help you locate items in a list.

Working with Python

The Python version of the code relies on lambda functions because the process is straightforward enough to avoid using multiple lines. Here is the Python code used for this example:

```python
findNext = lambda x, obj: -1 if len(obj) == 0 \
    or len(obj) == 1 \
    else obj[1] if x == obj[0] \
        else findNext(x, obj[1:])

findPrev = lambda x, obj: -1 if len(obj) == 0 \
    or len(obj) == 1 \
    else obj[0] if x == obj[1] \
        else findPrev(x, obj[1:])
```

LAMBDA TYPING IN PYTHON

You may wonder why the example code for the lambda functions in this chapter doesn't include data type as part of the argument list. The short answer is that you can't provide a data type according to the discussion at `https://www.python.org/dev/peps/pep-3107/#lambda`, which also provides reasons for this omission. If data typing is important, you need to use a function form of your lambda function.

Normally, you could place everything on a single line; this example uses line-continuation characters to accommodate this book's margins. As with the Haskell code, the example begins by verifying that the incoming list contains two or more values. It also returns `-1` when there is no next or previous value to find. The essential mechanism used, a comparison, is the same as the Haskell example. In this case, the Python code relies on slicing to reduce the size of the list on each pass. Figure 8-8 shows this example in action using both successful and unsuccessful searches.

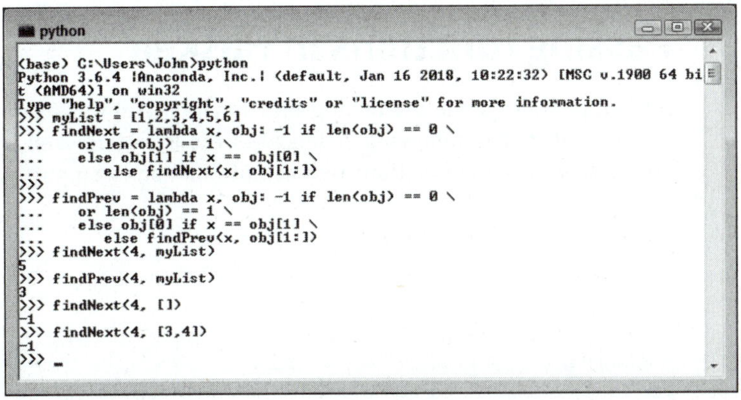

FIGURE 8-8: Python uses the same mechanism as Haskell to find matches.

Passing Functions Instead of Variables

This section looks at passing functions to other functions, using a technique that is one of the most powerful features of the functional programming paradigm. Even though this section isn't strictly about recursion, you can use the technique in it with recursion. The example code is simplified to make the principle clear.

Understanding when you need a function

Being able to pass a function to another function provides much needed flexibility. The passed function can modify the receiving function's response without modifying that receiving function's execution. The two functions work in tandem to create output that's an amalgamation of both.

TIP

Normally, when you use this function-within-a-function technique, one function determines the process used to produce an output, while the second function determines how the output is achieved. This isn't always the case, but when creating a function that receives another function as input, you need to have a particular goal in mind that actually requires that function as input. Given the complexity of debugging this sort of code, you need to achieve a specific level of flexibility by using a function rather than some other input.

Also tempting is to pass a function to another function to mask how a process works, but this approach can become a trap. Try to execute the function externally when possible and input the result instead. Otherwise, you might find yourself trying to discover the precise location of a problem, rather than processing data.

Passing functions in Haskell

Haskell provides some interesting functionality, and this example shines a light on some of it. The following code shows the use of Haskell signatures to good effect when creating functions that accept functions as input:

```
doAdd :: Int -> Int -> Int
doAdd x y = x + y

doSub :: Int -> Int -> Int
doSub x y = x - y

cmp100 :: (Int -> Int -> Int) -> Int -> Int -> Ordering
cmp100 f x y = compare 100 (f x y)
```

The signatures and code for both doAdd and doSub are straightforward. Both functions receive two integers as input and provide an integer as output. The first function simply adds to values; the second subtracts them. The signatures are important to the next step.

The second step is to create the cmp100 function, which accepts a function as the first input. Notice the (Int -> Int -> Int) part of the signature. This section indicates a function (because of the parentheses) that accepts two integers as input and provides an integer as output. The function in question can be any function that has these characteristics. The next part of the signature shows that the function will receive two integers as input (to pass along to the called function) and provide an order as output.

The actual code shows that the function is called with the two integers as input. Next, compare is called with 100 as the first value and the result of whatever happens in the called function as the second input. Figure 8-9 shows the example code in action. Notice that the two numeric input values provide different results, depending on which function you provide.

```
WinGHCi
File  Edit  Actions  Tools  Help

GHCi, version 8.2.2: http://www.haskell.org/ghc/  :? for help
Prelude> :load function.hs
[1 of 1] Compiling Main             ( function.hs, interpreted )
Ok, one module loaded.
*Main> doAdd 99 2
101
*Main> doSub 99 2
97
*Main> cmp100 doAdd 99 2
LT
*Main> cmp100 doSub 99 2
GT
*Main> cmp100 doAdd 99 1
EQ
*Main>
```

FIGURE 8-9:
Depending on the function you pass, the same numbers produce different results.

Passing functions in Python

For the example in this section, you can't use a lambda function to perform the required tasks with Python, so the following code relies on standard functions instead. Notice that the functions are the same as those provided in the previous example for Haskell and they work nearly the same way.

```python
def doAdd(x, y):
    return x + y

def doSub(x, y):
    return x - y

def compareWithHundred(function, x, y):
    z = function(x, y)
    out = lambda x: "GT" if 100 > x \
        else "EQ" if 100 == x else "LT"
    return out(z)
```

The one big difference is that Python doesn't provide a compare function that provides the same sort of output as the Haskell compare function. In this case, a lambda function performs the comparison and provides the proper output. Figure 8-10 shows this example in action.

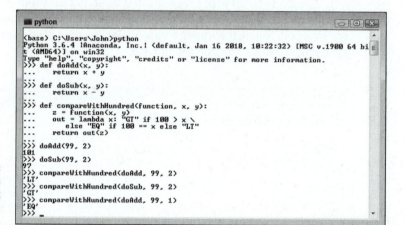

FIGURE 8-10: Python makes performing the comparison a bit harder.

Defining Common Recursion Errors

Recursion can actually cause quite a few problems — not because recursion is brittle or poorly designed, but because developers don't rely on it very often. In fact, most developers avoid using recursion because they see it as hard when it truly isn't. A properly designed recursive routine has an elegance and efficiency not found in other programming structures. With all this said, developers still do make some common errors when using recursion, and the following sections give you some examples.

Forgetting an ending

When working with looping structures, you can see the beginning and end of the loop with relative ease. Even so, most looping structures aren't foolproof when it comes to providing an exit strategy, and developers do make mistakes. Recursive routines are harder because you can't see the ending. All you really see is a function that calls itself, and you know where the beginning is because it's where the function is called initially.

Recursion doesn't rely on state, so you can't rely on it to perform a task a set number of times (as in a for loop) and then end unless you design the recursion to end in this manner. Likewise, even though recursion does seem like a while statement because it often relies on a condition to end, it doesn't use a variable, so there is nothing to update. Recursion does end when it detects an event or meets a condition, but the circumstances for doing so differ from loops. Consequently, you need to exercise caution when using recursion to ensure that the recursion will end before the host system runs out of stack space to support the recursive routine.

Passing data incorrectly

Each level of recursion generally requires some type of data input. Otherwise, knowing whether an event has occurred or a condition is met becomes impossible. The problem isn't with understanding the need to pass data, it's with understanding the need to pass the *correct* data. When most developers write applications, the focus is on the current level — that is, where the application is at now. However, when working through a recursion problem, the focus is instead on where the application will be in the future — the next step, anyway. This ability to write code for the present but work with data in the future can make it difficult to understand precisely what to pass to the next level when the function calls itself again.

Likewise, when processing the data that the previous level passed, see the data in the present is often difficult. When developers write most applications, they look at the past. For example, when looking at user input, the developer sees that input as the key (or keys) the user pressed — past tense. The data received from the previous level in a recursion is the present, which can affect how you view that data when writing code.

Defining a correct base instruction

Recursion is all about breaking down a complex task into one simple task. The complex task, such as processing a list, looks impossible. So, what you do is think about what is possible. You need to consider the essential output for a single data element and then use that as your base instruction. For example, when processing a list, you might simply want to display the data element's value. That instruction becomes your base instruction. Many recursion processes fail because the developer looks at the wrong end first. You need to think about the conclusion of the task first, and then work on progressively more complex parts of the data after that.

REMEMBER

Simpler is always better when working with recursion. The smaller and simpler your base instruction, the smaller and simpler the rest of the recursion tends to become. A base instruction shouldn't include any sort of logic or looping. In fact, if you can get it down to a single instruction, that's the best way to go. You don't want to do anything more than absolutely necessary when you get to the base instruction.

Chapter **9**

Advancing with Higher-Order Functions

Previous chapters in this book spend a lot of time looking at how to perform basic application tasks and viewing data to see what it contains in various ways. Just viewing the data won't do you much good, however. Data rarely comes in the form you need it and even if it does, you still want the option to mix it with other data to create yet newer views of the real world. Gaining the ability to shape data in certain ways, throw out what you don't need, refine its appearance, change its type, and otherwise condition it to meet your needs is the essential goal of this chapter.

Shaping, in the form of slicing and dicing, is the most common kind of manipulation. Data analysis can take hours, days, or even weeks at times. Anything you can do to refine the data to match specific criteria is important in getting answers fast. Obtaining answers quickly is essential in today's world. Yes, you need the correct answer, but if someone else gets the correct answer first, you may find that the answer no longer matters. You lose your competitive edge.

Also essential is having the right data. The use of data mapping enables you to correlate data between information systems so that you can draw new conclusions. In addition, information overload, especially the wrong kind of information, is never productive, so filtering is essential as well. The combination of mapping and filtering lets you control the dataset content without changing the dataset truthfulness. In short, you get a new view of the same old information.

Data presentation — that is, its organization — is also important. The final section of this chapter discusses the issue of how to organize data to better see the patterns it contains. Given that there isn't just one way to organize data, one presentation may show one set of patterns, and another presentation could display other patterns. The goal of all this data manipulation is to see something in the data that you haven't seen before. Perhaps the data will give you an idea for a new product or help you market products to a new group of users. The possibilities are nearly limitless.

Considering Types of Data Manipulation

When you mention the term data manipulation, you convey different information to different people, depending on their particular specialty. An overview of data manipulation may include the term *CRUD*, which stands for Create, Read, Update, and Delete. A database manager may view data solely from this low-level perspective that involves just the mechanics of working with data. However, a database full of data, even accurate and informative data, isn't particularly useful, even if you have all the best CRUD procedures and policies in place. Consequently, just defining data manipulation as CRUD isn't enough, but it's a start.

WARNING

To make really huge datasets useful, you must transform them in some manner. Again, depending on whom you talk to, transformation can take on all sorts of meanings. The one meaning that you won't see in this book is the modification of data such that it implies one thing when it actually said something else at the outset (think of this as spin doctoring the data). In fact, it's a good idea to avoid this sort of data manipulation entirely because you can end up with completely unpredictable results when performing analysis, even if those results initially look promising and even say what you feel they should say.

Another kind of data transformation actually does something worthwhile. In this case, the meaning of the data doesn't change; only the presentation of the data changes. You can separate this kind of transformation into a number of methods that include (but aren't necessarily limited to) tasks such as the following:

>> **Cleaning:** As with anything else, data gets dirty. You may find that some of it is missing information and some of it may actually be correct but outdated. In fact, data becomes dirty in many ways, and you always need to clean it before you can use it. *Machine Learning For Dummies,* by John Paul Mueller and Luca Massaron (Wiley), discusses the topic of cleaning in considerable detail.

>> **Verification:** Establishing that data is clean doesn't mean that the data is correct. A dataset may contain many entries that seem correct but really aren't. For example, a birthday may be in the right form and appear to be correct until you determine that the person in question is more than 200 years old. A part

number may appear in the correct form, but after checking, you find that your organization never produced a part with that number. The act of verification helps ensure the veracity of any analysis you perform and generates fewer outliers to skew the results.

>> **Data typing:** Data can appear to be correct and you can verify it as true, yet it may still not work. A significant problem with data is that the type may be incorrect or it may appear in the wrong form. For example, one dataset may use integers for a particular column (feature), while another uses floating-point values for the same column. Likewise, some datasets may use local time for dates and times, while others might use GMT. The transformation of the data from various datasets to match is an essential task, yet the transformation doesn't actually change the data's meaning.

>> **Form:** Datasets come with many form issues. For example, one dataset may use a single column for people's names, while another might use three columns (first, middle, and last), and another might use five columns (prefix, first, middle, last, and suffix). The three datasets are correct, but the form of the information is different, so a transformation is needed to make them work together.

>> **Range:** Some data is categorical or uses specific ranges to denote certain conditions. For example, probabilities range from 0 to 1. In some cases, there isn't an agreed-upon range. Consequently, you find data appearing in different ranges even though the data refers to the same sort of information. Transforming all the data to match the same range enables you to perform analysis by using data from multiple datasets.

>> **Baseline:** You hear many people talk about dB when considering audio output in various scenarios. However, a decibel is simply a logarithmic ratio, as described at http://www.animations.physics.unsw.edu.au/jw/dB.htm. Without a reference value or a baseline, determining what the dB value truly means is impossible. For audio, the dB is referenced to 1 volt (dBV), as described at http://www.sengpielaudio.com/calculator-db-volt.htm. The reference is standard and therefore implied, even though few people actually know that a reference is involved. Now, imagine the chaos that would result if some people used 1 volt for a reference and others used 2 volts. dBV would become meaningless as a unit of measure. Many kinds of data form a ratio or other value that requires a reference. Transformations can adjust the reference or baseline value as needed so that the values can be compared in a meaningful way.

REMEMBER

You can come up with many other transformations. The point of this section is that the method used determines the kind of transformation that occurs, and you must perform certain kinds of transformations to make data useful. Applying an incorrect transformation or the correct transformation in the wrong way will result in useless output even when the data itself is correct.

Performing Slicing and Dicing

Slicing and dicing are two ways to control the size of a dataset. *Slicing* occurs when you use a subset of the dataset in a single axis. For example, you may want only certain records (also called cases) or you may want only certain columns (also called features). *Dicing* occurs when you perform slicing in multiple directions. When working with two-dimensional data, you select certain rows and certain columns from those rows. You see dicing used more often with three-dimensional or higher data, when you want to restrict the x-axis and the y-axis but keep all the z-axis (as an example). The following sections describe slicing and dicing in more detail and demonstrate how to perform this task using both Haskell and Python.

Keeping datasets controlled

Datasets can become immense. The data continues to accumulate from various sources until it becomes impossible for the typical human to comprehend it all. So slicing and dicing might at first seem to be a means for making data more comprehensible. It can do that, but making the data comprehensible isn't the point. Too much data can even overwhelm a computer — not in the same way as a human gets overwhelmed, because a computer doesn't understand anything, but to the point where processing proceeds at a glacial pace. As the cliché says, time is money, which is precisely why you want to control dataset size. The more focused you can make any data analysis, the faster the analysis will proceed.

REAL-WORLD SLICING AND DICING

The examples in this chapter are meant to demonstrate techniques used with the functional programming paradigm in the simplest manner possible. With this in mind, the examples rely on native language capabilities whenever possible. In the real world, when working with large applications rather than experimenting, you use libraries to make the task easier — especially when working with immense datasets. For example, Python developers often rely on NumPy (http://www.numpy.org/) or pandas (https://pandas.pydata.org/) when performing this task. Likewise, Haskell developers often use hmatrix (https://hackage.haskell.org/package/hmatrix), repa (https://hackage.haskell.org/package/repa), and vector (https://hackage.haskell.org/package/vector) to perform the same tasks. The libraries vary in functionality, provide language-specific features, and make it tough to compare code. Consequently, when you're initially discovering how to perform a technique, it's often best to rely on native capability and then add library functionality to augment the language.

Sometimes you must use slicing and dicing to break the data down into training and testing units for computer technologies such as machine learning. You use the training set to help an algorithm perform the correct processing in the correct way through examples. The testing set then verifies that the training went as planned. Even though machine learning is the most prominent technology today that requires breaking data into groups, you can find other examples. Many database managers work better when you break data into pieces and perform batch processing on it, for example.

WARNING

Slicing and dicing can give you a result that doesn't actually reflect the realities of the data as a whole. If the data isn't randomized, one piece of the data may contain more of some items than the other piece. Consequently, you must sometimes randomize (shuffle) the dataset before using slicing and dicing techniques on it.

Focusing on specific data

Slicing and dicing techniques can also help you improve the focus of a particular analysis. For example, you may not actually require all the columns (features) in a dataset. Removing the extraneous columns can actually make the data easier to use and provide results that are more reliable.

Likewise, you may need to remove unneeded information from the dataset. For example, a dataset that contains entries from the last three years requires slicing or dicing when you need to analyze only the results from one year. Even though you could use various techniques to ignore the extra entries in code, eliminating the unwanted years from the dataset using slicing and dicing techniques makes more sense.

TIP

Be sure to keep slicing and dicing separate from filtering. Slicing and dicing focuses on groups of randomized data for which you don't need to consider individual data values. Slicing out a particular year from a dataset containing sales figures is different from filtering the sales produced by a particular agent. Filtering looks for specific data values regardless of which group contains that value. The "Filtering Data" section, later in this chapter, discusses filtering in more detail, but just keep in mind that the two techniques are different.

Slicing and dicing with Haskell

Haskell slicing and dicing requires a bit of expertise to understand because you don't directly access the slice as you might with other languages through indexing.

Of course, there are libraries that encapsulate the process, but this section reviews a native language technique that will do the job for you using the take and drop functions. Slicing can be a single-step process if you have the correct code. To begin, the following code begins with a one-dimensional list, let myList = [1, 2, 3, 4, 5].

```
-- Display the first two elements.
take 2 myList


-- Display the remaining three elements.
drop 2 myList


-- Display a data slice of just the center element.
take 1 $ drop 2 myList
```

The slice created by the last statement begins by dropping the first two elements using drop 2 myList, leaving [3,4,5]. The $ operator connects this output to the next function call, take 1, which produces an output of [3]. Using this little experiment, you can easily create a slice function that looks like this:

```
slice xs x y = take y $ drop x xs
```

To obtain just the center element from myList, you would call slice myList 2 1, where 2 is the zero-based starting index and 1 is the length of the output you want. Figure 9-1 shows how this sequence works.

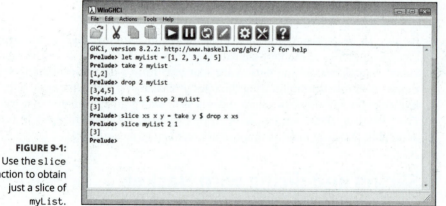

FIGURE 9-1:
Use the slice function to obtain just a slice of myList.

Of course, slicing that works only on one-dimensional arrays isn't particularly useful. You can test the `slice` function on a two-dimensional array by starting with a new list, `let myList2 = [[1,2],[3,4],[5,6],[7,8],[9,10]]`. Try the same call as before, `slice myList2 2 1`, and you see the expected output of `[[5,6]]`. So, `slice` works fine even with a two-dimensional list.

Dicing is somewhat the same, but not quite. To test the `dice` function, begin with a slightly more robust list, `let myList3 = [[1,2,3],[4,5,6],[7,8,9], [10,11,12],[13,14,15]]`. Because you're now dealing with the inner values rather than the lists contained with a list, you must rely on recursion to perform the task. The "Defining the need for repetition" section of Chapter 8 introduces you to the `forM` function, which repeats a particular code segment. The following code shows a simplified, but complete, dicing sequence.

```
import Control.Monad
let myList3 =
    [[1,2,3],[4,5,6],[7,8,9],[10,11,12],[13,14,15]]
slice xs x y = take y $ drop x xs
dice lst x y = forM lst (\i -> do return(slice i x y))
lstr = slice myList3 1 3
lstr
lstc = dice lstr 1 1
lstc
```

To use `forM`, you must `import Control.Monad`. The `slice` function is the same as before, but you must define it within the scope created after the import. The `dice` function uses `forM` to examine every element within the input list and then slice it as required. What you're doing is slicing the list within the list. The next items of code first slice `myList3` into rows, and then into columns. The output is as you would expect: `[[5],[8],[11]]`. Figure 9-2 shows the sequence of events.

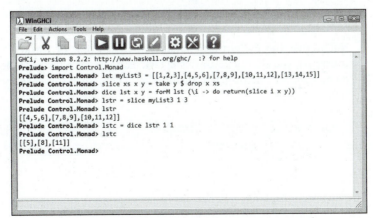

FIGURE 9-2:
Dicing is a two-step process.

Slicing and dicing with Python

In some respects, slicing and dicing is considerably easier in Python than in Haskell. For one thing, you use indexes to perform the task. Also, Python offers more built-in functionality. Consequently, the one-dimensional list example looks like this:

```
myList = [1, 2, 3, 4, 5]

print(myList[:2])
print(myList[2:])
print(myList[2:3])
```

The use of indexes enables you to write the code succinctly and without using special functions. The output is as you would expect:

```
[1, 2]
[3, 4, 5]
[3]
```

Slicing a two-dimensional list is every bit as easy as working with a one-dimensional list. Here's the code and output for the two-dimensional part of the example:

```
myList2 = [[1,2],[3,4],[5,6],[7,8],[9,10]]

print(myList2[:2])
print(myList2[2:])
print(myList2[2:3])

[[1, 2], [3, 4]]
[[5, 6], [7, 8], [9, 10]]
[[5, 6]]
```

Notice that the Python functionality matches that of Haskell's take and drop functions; you simply perform the task using indexes instead. Dicing does require using a special function, but the function is concise in this case and doesn't require multiple steps:

```
def dice(lst, rb, re, cb, ce):
    lstr = lst[rb:re]
    lstc = []
```

```
    for i in lstr:
        lstc.append(i[cb:ce])
    return lstc
```

In this case, you can't really use a lambda function — or not easily, at least. The code slices the incoming list first and then dices it, just as in the Haskell example, but everything occurs within a single function. Notice that Python requires the use of looping, but this function uses a standard `for` loop instead of relying on recursion. The disadvantage of this approach is that the loop relies on state, which means that you can't really use it in a fully functional setting. Here's the test code for the dicing part of the example:

```
myList3 = [[1,2,3],[4,5,6],[7,8,9],[10,11,12],[13,14,15]]

print(dice(myList3, 1, 4, 1, 2))

[[5], [8], [11]]
```

Mapping Your Data

You can find a number of extremely confusing references to the term *map* in computer science. For example, a map is associated with database management (see `https://en.wikipedia.org/wiki/Data_mapping`), in which data elements are mapped between two distinct data models. However, for this chapter, *mapping* refers to a process of applying a high-order function to each member of a list. Because the function is applied to every member of the list, the relationships among list members is unchanged. Many reasons exist to perform mapping, such as ensuring that the range of the data falls within certain limits. The following sections of the chapter help you better understand the uses for mapping and demonstrate the technique using the two languages supported in this book.

Understanding the purpose of mapping

The main idea behind mapping is to apply a function to all members of a list or similar structure. Using mapping can help you adjust the range of the values or prepare the values for particular kinds of analysis. Functional languages originated the idea of mapping, but mapping now sees use in most programming languages that support first-class functions.

REMEMBER

The goal of mapping is to apply the function or functions to a series of numbers equally to achieve specific results. For example, squaring the numbers can rid the series of any negative values. Of course, you can just as easily take the absolute value of each number. You may need to convert a probability between 0 and 1 to a percentage between 0 and 100 for a report or other output. The relationship between the values will stay the same, but the range won't. Mapping enables you to obtain specific data views.

Performing mapping tasks with Haskell

Haskell is one of the few computer languages whose `map` function isn't necessarily what you want. For example, the `map` associated with `Data.Map.Strict`, `Data.Map.Lazy`, and `Data.IntMap` works with the creation and management of dictionaries, not the application of a consistent function to all members of a list (see `https://haskell-containers.readthedocs.io/en/latest/map.html` and `http://hackage.haskell.org/package/containers-0.5.11.0/docs/Data-Map-Strict.html` for details). What you want instead is the `map` function that appears as part of the base prelude so that you can access `map` without importing any libraries.

The `map` function accepts a function as input, along with one or more values in a list. You might create a function, `square`, that outputs the square of the input value: `square x = x * x`. A list of values, `items = [0, 1, 2, 3, 4]`, serves as input. Calling `map square items` produces an output of `[0,1,4,9,16]`. Of course, you could easily create another function: `double x = x + x`, with a `map double items` output of `[0,2,4,6,8]`. The output you receive clearly depends on the function you use as input (as expected).

TIP

You can easily get overwhelmed trying to create complex functions to modify the values in a list. Fortunately, you can use the composition operator (., or dot) to combine them. Haskell actually applies the second function first. Consequently, `map (square.double) items` produces an output of `[0,4,16,36,64]` because Haskell doubles the numbers first, and then squares them. Likewise, `map (double.square) items` produces an output of `[0,2,8,18,32]` because squaring occurs first, followed by doubling.

The apply operator ($) is also important to mapping. You can create a condition for which you apply an argument to a list of functions. As shown in Figure 9-3, you place the argument first in the list, followed by the function list (`map ($4) [double, square]`). The output is a list with one element for each function, which is `[8,16]` in this case. Using recursion would allow you to apply a list of numbers to a list of functions.

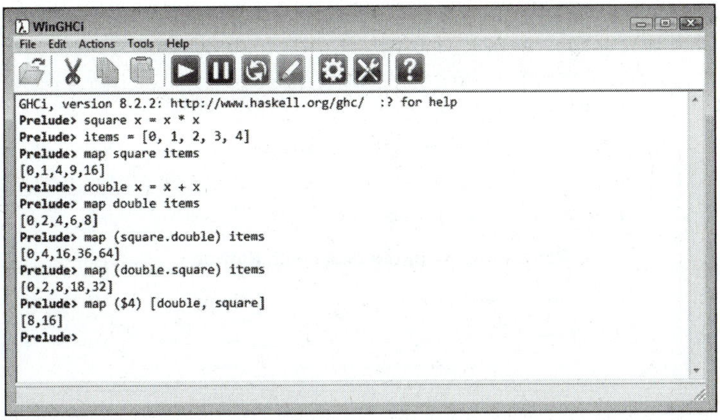

FIGURE 9-3:
You can apply a
single value to a
list of functions.

```
WinGHCi
File  Edit  Actions  Tools  Help

GHCi, version 8.2.2: http://www.haskell.org/ghc/  :? for help
Prelude> square x = x * x
Prelude> items = [0, 1, 2, 3, 4]
Prelude> map square items
[0,1,4,9,16]
Prelude> double x = x + x
Prelude> map double items
[0,2,4,6,8]
Prelude> map (square.double) items
[0,4,16,36,64]
Prelude> map (double.square) items
[0,2,8,18,32]
Prelude> map ($4) [double, square]
[8,16]
Prelude>
```

Performing mapping tasks with Python

Python performs many of the same mapping tasks as Haskell, but often in a slightly different manner. Look, for example, at the following code:

```python
square = lambda x: x**2
double = lambda x: x + x
items = [0, 1, 2, 3, 4]

print(list(map(square, items)))
print(list(map(double, items)))
```

You obtain the same output as you would with Haskell using similar code. However, note that you must convert the map object to a list object before printing it. Given that Python is an impure language, creating code that processes a list of inputs against two or more functions is relatively easy, as shown in this code:

```python
funcs = [square, double]

for i in items:
    value = list(map(lambda items: items(i), funcs))
    print(value)
```

Note that, as with the Haskell code, you're actually applying individual list values against the list of functions. However, Python requires a lambda function to get the job done. Figure 9-4 shows the output from the example.

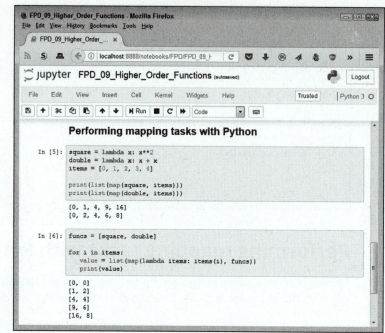

FIGURE 9-4:
Using multiple paradigms in Python makes mapping tasks easier.

The figure shows a Jupyter notebook with the following content:

Performing mapping tasks with Python

```python
square = lambda x: x**2
double = lambda x: x + x
items = [0, 1, 2, 3, 4]

print(list(map(square, items)))
print(list(map(double, items)))
```

```
[0, 1, 4, 9, 16]
[0, 2, 4, 6, 8]
```

```python
funcs = [square, double]

for i in items:
    value = list(map(lambda items: items(i), funcs))
    print(value)
```

```
[0, 0]
[1, 2]
[4, 4]
[9, 6]
[16, 8]
```

Filtering Data

Most programming languages provide specialized functions for filtering data today. Even when the language doesn't provide a specialized function, you can use common methods to perform filtering manually. The following sections discuss what filtering is all about and how to use the two target languages to perform the task.

Understanding the purpose of filtering

Data filtering is an essential tool in removing outliers from datasets, as well as selecting specific data based on one or more criteria for analysis. While slicing and dicing selects data regardless of specific content, data filtering makes specific selections to achieve particular goals. Consequently, the two techniques aren't mutually exclusive; you may well employ both on the same dataset in an effort to locate the particular data needed for an analysis. The following sections discuss details of filtering use and provide examples of simple data filtering techniques for both of the languages used in this book.

Developers often apply slicing and dicing, mapping, and filtering together to shape data in a manner that doesn't change the inherent relationships among data elements. In all three cases, the data's organization remains unchanged, and an element that is twice the size of another element tends to remain in that same relationship. Modifying data range, the number of data elements, and other factors in a dataset that don't modify the data's content — its relationship to the environment from which it was taken — is common in data science in preparation for performing tasks such as analysis and comparison, along with creating single, huge datasets from numerous smaller datasets. Filtering enables you to ensure that the right data is in the right place and at the right time.

Using Haskell to filter data

Haskell relies on a `filter` function to remove unwanted elements from lists and other dataset structures. The `filter` function accepts two inputs: a description of what you want removed and the list of elements to filter. The filter descriptions come in three forms:

» Special keywords, such as odd and even

» Simple logical comparisons, such as ›

» Lambda functions, such as \x –> mod x 3 == 0

To see how this all works, you could create a list such as items = [0, 1, 2, 3, 4, 5]. Figure 9-5 shows the results of each of the filtering scenarios.

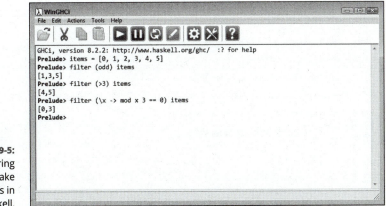

FIGURE 9-5:
Filtering descriptions take three forms in Haskell.

WARNING

You want to carefully consider the use of Haskell operators when performing any task, but especially filtering. For example, at first look, rem and mod might not seem much different. Using rem 5 3 produces the same output as mod 5 3 (an output of 2). However, as noted at https://stackoverflow.com/questions/5891140/difference-between-mod-and-rem-in-haskell, a difference arises when working with a negative number. In this situation, mod 3 (-5) produces an output of -2, while rem 3 (-5) produces an output of 3.

Using Python to filter data

Python doesn't provide a few of the niceties that Haskell does when it comes to filtering. For example, you don't have access to special keywords, such as odd or even. In fact, all the filtering in Python requires the use of lambda functions. Consequently, to obtain the same results for the three cases in the previous section, you use code like this:

```python
items = [0, 1, 2, 3, 4, 5]

print(list(filter(lambda x: x % 2 == 1, items)))
print(list(filter(lambda x: x > 3, items)))
print(list(filter(lambda x: x % 3 == 0, items)))
```

Notice that you must convert the filter output using a function such as list. You don't have to use list; you could use any data structure, including set and tuple. The lambda function you create must evaluate to True or False, just as it must with Haskell. Figure 9-6 shows the output from this example.

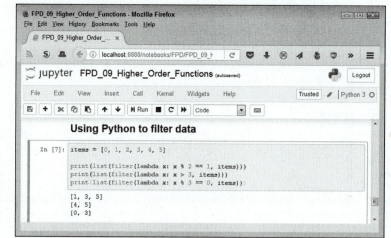

FIGURE 9-6: Python lacks some of the Haskell special filtering features.

Organizing Data

None of the techniques discussed so far changes the organization of the data directly. All these techniques can indirectly change organization through a process of data selection, but that's not the goal of the methods applied. However, sometimes you do need to change the organization of the data. For example, you might need it sorted or grouped based on specific criteria. In some cases, organizing the data can also mean to randomize it in some manner to ensure that an analysis reflects the real world. The following sections discuss the kinds of organization that most people apply to data; also covered is how you can implement sorting using the two languages that appear in this book.

Considering the types of organization

Organization — the forming of any object based on a particular pattern—is an essential part of working with data for humans. The coordination of elements within a dataset based on a particular need is usually the last step in making the data useful, except when other parts of the cleaning process require organization to work properly. How something is organized affects the way in which humans view it, and organizing the object in some other manner will change the human perspective, so often people find themselves organizing datasets one way and then reorganizing them in another. No right or wrong way to organize data exists; you just want to use the approach that works best for viewing the information in a way that helps see the desired pattern.

TIP

You can think about organization in a number of ways. Sometimes the best organization is disorganization. Seeing patterns in seemingly random patterns finds a place in many areas of life, including art (see the stereograms at http://www.vision3d.com/sghidden.html as an example). A pattern is what you make of it, so sometimes thinking about what you want to see, rather than making things neat and tidy, is the best way to achieve your objectives. The following list provides some ideas on organization, most of which you have thought about, but some of which you likely haven't. The list is by no means exhaustive.

>> **Sorting:** One of the most common ways to organize data is to sort it, with the alphanumeric sort being the most common. However, sorts need not be limited to ordering the data by the alphabet or computer character number. For example, you could sort according to value length or by commonality. In fact, the idea of sorting simply means placing the values in an order from greatest to least (or vice versa) according to whatever criteria the sorter deems necessary.

>> **Grouping:** Clustering data such that the data with the highest degree of commonality is together is another kind of sorting. For example, you might group data by value range, with each range forming a particular group. As with sorting, grouping criteria can be anything. You might choose to group textual data by the number of vowels contained in each element. You might group numeric data according to an algorithm of some sort. Perhaps you want all the values that are divisible by 3 in one bin and those that are divisible by 7 in another, with a third bin holding values that can't be divided by either.

>> **Categorizing:** Analyzing the data and placing data that has the same properties together is another method of organization. The properties can be anything. Perhaps you need to find values that match specific colors, or words that impart a particular kind of meaning. The values need not hold any particular commonality; they just need to have the same properties.

>> **Shuffling:** Disorganization can be a kind of organization. Chaos theory (see `https://fractalfoundation.org/resources/what-is-chaos-theory/` for an explanation) finds use in a wide variety of everyday events. In fact, many of today's sciences rely heavily on the effects of chaos. Data shuffling often enhances the output of algorithms and creates conditions that enable you to see unexpected patterns. Creating a kind of organization through the randomization of data may seem counter to human thought, but it works nonetheless.

Sorting data with Haskell

Haskell provides a wide variety of sorting mechanisms, such that you probably won't have to resort to doing anything of a custom nature unless your data is unique and your requirements are unusual. However, getting the native functionality that's found in existing libraries can prove a little daunting at times unless you think the process through first. To start, you need a list that's a little more complex than others used in this chapter: `original = [(1, "Hello"), (4, "Yellow"), (5, "Goodbye"), (2, "Yes"), (3, "No")]`. Use the following code to perform a basic sort:

```
import Data.List as Dl
sort original
```

The output is based on the first member of each tuple: `[(1,"Hello"),(2,"Yes"), (3,"No"),(4,"Yellow"),(5,"Goodbye")]`. If you want to perform a reverse sort, you can use the following call instead:

```
(reverse . sort) original
```

REMEMBER

Notice how the reverse and sort function calls appear in this example. You must also include the space shown between reverse, sort, and the composition operator (.). The problem with using this approach is that Haskell must go through the list twice: once to sort it and once to reverse it. An alternative is to use the sortBy function, as shown here:

```
sortBy (\x y -> compare y x) original
```

The sortBy function lets you use any comparison function needed to obtain the desired result. For example, you might not be interested in sorting by the first member of the tuple but instead prefer to sort by the second member. In this case, you must use the snd function from Data.Tuple (which loads with Prelude) with the comparing function from Data.Ord (which you must import), as shown here:

```
import Data.Ord as Do
sortBy (comparing $ snd) original
```

REMEMBER

Notice how the call applies comparing to snd using the apply operator ($). Using the correct operator is essential to make sorts work. The results are as you would expect: [(5,"Goodbye"),(1,"Hello"),(3,"No"),(4,"Yellow"),(2,"Yes")]. However, you might not want a straight sort. You really may want to sort by the length of the words in the second member of the tuple. In this case, you can make the following call:

```
sortBy (comparing $ length . snd) original
```

The call applies comparing to the result of the composition of snd, followed by length (essentially, the length of the second tuple member). The output reflects the change in comparison: [(3,"No"),(2,"Yes"),(1,"Hello"),(4,"Yellow"), (5,"Goodbye")]. The point is that you can sort in any manner needed using relatively simple statements in Haskell unless you work with complex data.

Sorting data with Python

The examples in this section use the same list as that found in the previous section: original = [(1, "Hello"), (4, "Yellow"), (5, "Goodbye"), (2, "Yes"), (3, "No")], and you'll see essentially the same sorts, but from a Python perspective. To understand these examples, you need to know how to use the sort method, versus the sorted function. When you use the sort method, Python changes the original list, which may not be what you want. In addition, sort works only with lists, while `sorted` works with any iterable. The `sorted` function produces output that doesn't change the original list. Consequently, if you want to maintain your original list form, you use the following call:

```
sorted(original)
```

The output is sorted by the first member of the tuple: `[(1, 'Hello'), (2, 'Yes'), (3, 'No'), (4, 'Yellow'), (5, 'Goodbye')]`, but the original list remains intact. Reversing a list requires the use of the `reverse` keyword, as shown here:

```
sorted(original, reverse=True)
```

Both Haskell and Python make use of lambda functions to perform special sorts. For example, to sort by the second element of the tuple, you use the following code:

```
sorted(original, key=lambda x: x[1])
```

TIP

The `key` keyword is extremely flexible. You can use it in several ways. For example, `key=str.lower` would perform a case-insensitive sort. Some of the common lambda functions appear in the `operator` module. For example, you could also sort by the second element of the tuple using this code:

```
from operator import itemgetter
sorted(original, key=itemgetter(1))
```

You can also create complex sorts. For example, you can sort by the length of the second tuple element by using this code:

```
sorted(original, key=lambda x: len(x[1]))
```

REMEMBER

Notice that you must use a lambda function when performing a custom sort. For example, trying this code will result in an error:

```
sorted(original, key=len(itemgetter(1)))
```

Even though `itemgetter` is obtaining the key from the second element of the tuple, it doesn't possess a length. To use the second tuple's length, you must work with the tuple directly.

Chapter **10**

Dealing with Types

The term *type* takes on new meaning when working with functional languages. In other languages, when you speak of a type, you mean the label attached to a certain kind of data. This label tells the compiler how to interact with the data. The label is intimately involved with the value. In functional languages, type is more about mapping. You compose functions that express a mapping of or transformation between types of data. The function is a mathematical expression that defines the transformation using a representation of the math involved in the transformation. Just how a language supports this idea of mapping and transformation depends on how it treats underlying types. Because Haskell actually provides a purer approach with regard to type and the functional programming paradigm, this chapter focuses a little heavier on Haskell.

As with other languages, you can create new types as needed in functional languages. However, the manner in which you create and use new types differs because of how you view type. Interestingly enough, creating new types can be easier in functional languages because the process is relatively straightforward and the result is easier to read in most cases.

The other side of the coin is that functional languages tend toward stricter management of type. (This is true for the most part, at least. Exceptions definitely exist, such as JavaScript, which is being fixed; see `https://www.w3schools.com/js/js_strict.asp` for details.) Because of this strictness, you need to know how to understand, manage, and fix type errors. In addition, you should understand how the use of type affects issues such as missing data. The chapter includes examples in both Haskell and Python to demonstrate all of the various aspects of type.

Developing Basic Types

Functional languages provide a number of methods for defining type. Remember that no matter what programming paradigm you use, the computer sees numbers — 0s and 1s, actually. The concept of type has no meaning for the computer; type is there to help the humans writing the code. As with anything, when working with types, starting simply is best. The following sections examine the basics of type in the functional setting and discuss how to augment those types to create new types.

Understanding the functional perception of type

As mentioned in the introduction, a pure functional language, such as Haskell, uses expressions for everything. Because everything is an expression, you can substitute functions that provide the correct output in place of a value. However, values are also expressions, and you can test this idea by using :t to see their types. When you type :t **True** and press Enter, you see True :: Bool as output because True is an expression that produces a Bool output. Likewise, when you type :t **5 == 6** and press Enter, you see 5 == 6 :: Bool as the output. Any time you use the :t command, you see the definition of the type of whatever you place after the command.

Python takes a similar view, but in a different manner, because it supports multiple programming paradigms. In Python, you point to an object using a name. The object contains the value and provides its associated properties. The object controls its use because it knows how to be that particular object. You can point to a different object using the name you define, but the original object remains unchanged. To see this perception of type, you use the Python type function. When you type **type(1)**, you see <class 'int'> as output. Other languages might say that the type of a value 1 is an int, rather than say that the type of a value 1 is an instance of the class int. If you create a variable by **typing myInt = 1** and pressing Enter, then use the type(myInt) function, you still see <class 'int'> as output. The name myInt merely points to an object that is an instance of class int. Even expressions work this way. For example, when you type **myAdd = 1 + 1** and then use type(myAdd), you still get <class 'int'> as output.

Considering the type signature

A number of nonfunctional languages use type signatures to good effect, although they may have slightly different names and slightly different uses, such as the function signature in C++. Even so, signatures used to describe the inputs and outputs of the major units of application construction for a language are nothing new. The type signature in Haskell is straightforward. You use one for the findNext function in Chapter 8:

```
findNext :: Int -> [Int] -> Int
```

In this case, the expression findNext (on the left side of the double colon) expects an Int and an [Int] (list) as input, and provides an Int as output. A type signature encompasses everything needed to fully describe an expression and helps relieve potential ambiguity concerning the use of the expression. Haskell doesn't always require that you provide a type signature (many of the examples in this book don't use one), but will raise an error if ambiguity exists in the use of an expression and you don't provide the required type signature. When you don't provide a type signature, the compiler infers one (as described in the previous section). Later sections of this chapter discuss some of the complexities of using type signatures.

Python can also use type signatures, but the philosophy behind Python is different from that of many other languages. The type signature isn't enforced by the interpreter, but IDEs and other tools can use the type signature to help you locate potential problems with your code. Consider this function with the type signature:

```
def doAdd (value1 : int, value2 : int) -> int:
    return value1 + value2
```

WARNING

The function works much as you might expect. For example, doAdd(1, 2) produces an output of 3. When you type **type((doAdd(1, 2)))** and press Enter, you also obtain the expected result of <class 'int'>. However, the philosophy of Python is that function calls will respect the typing needed to make the function work, so the interpreter doesn't perform any checks. The call doAdd("Hello", " Goodbye") produces an output of 'Hello Goodbye', which is most definitely not an int. When you type **type((doAdd("Hello", " Goodbye")))** and press Enter, you obtain the correct, but not expected, output of <class 'str'>.

One way around this problem is to use a static type checker such as mypy (http://mypy-lang.org/). When you call on this tool, it checks your code against the signature you provide.

TECHNICAL STUFF

A more complete type signature for Python would tend to include some sort of error trapping. In addition, you could use default values to make the intended input more apparent. For example, you could change doAdd to look like this:

```
def doAdd (value1 : int = 0, value2 : int = 0) -> int:
    if not isinstance(value1, int) or \
        not isinstance(value2, int):
            raise TypeError
    return value1 + value2
```

The problem with this approach is that it runs counter to the Python way of performing tasks. When you add type checking code of this sort, you automatically limit the potential for other people to use functions in useful, unexpected, and completely safe ways. Python relies on an approach called Duck Typing (see `http://wiki.c2.com/?DuckTyping` and `https://en.wikipedia.org/wiki/Duck_typing` for details). Essentially, if it walks like a duck and talks like a duck, it must be a duck, despite the fact that the originator didn't envision it as a duck.

Creating types

At some point, the built-in types for any language won't satisfy your needs and you'll need to create a custom type. The method used to create custom types varies by language. As noted in the "Understanding the functional perception of type" section, earlier in this chapter, Python views everything as an object. In this respect, Python is an object-oriented language within limits (for example, Python doesn't actually support data hiding). With this in mind, to create a new type in Python, you create a new class, as described at `https://docs.python.org/3/tutorial/classes.html` and `https://www.learnpython.org/en/Classes_and_Objects`. This book doesn't discuss object orientation to any degree, so you won't see much with regard to creating custom Python types.

Haskell takes an entirely different approach to the process that is naturally in line with functional programming principles. In fact, you may be amazed to discover the sorts of things you can do with very little code. The following sections offer an overview of creating types in Haskell, emphasizing the functional programming paradigm functionality.

Using AND

Haskell has this concept of adding types together to create a new kind of type. One of the operations you can perform on these types is AND, which equates to this type and this type as a single new type. In this case, you provide a definition like this one shown here.

```
data CompNum = Comp Int Int
```

It's essential to track the left and right side of the definition separately. The left side is the type constructor and begins with the `data` keyword. For now, you create a type constructor simply by providing a name, which is `CompNum` (for complex number, see `https://www.mathsisfun.com/numbers/complex-numbers.html` for details).

The right side is the data constructor. It defines the essence of the data type. In this case, it includes an identifier, Comp, followed by two Int values (the real component and the imaginary component). To create and test this type, you would use the following code:

```
x = Comp 5 7
:t x
```

The output, as you might expect, is x :: CompNum, and the new data type shows the correct data constructor. This particular version of CompNum has a problem. Type **x** by itself and you see the error message shown in Figure 10-1.

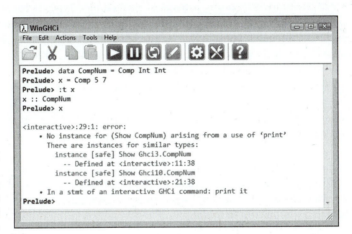

To fix this problem, you must tell the data type to derive the required functionality. The declarative nature of Haskell means that you don't actually have to provide an implementation; declaring that a data type does something is enough to create the implementation, as shown here:

```
data CompNum = Comp Int Int deriving Show
x = Comp 5 7
:t x
x
```

REMEMBER

The deriving keyword is important to remember because it makes your life much simpler. The new data type now works as expected (see Figure 10-2).

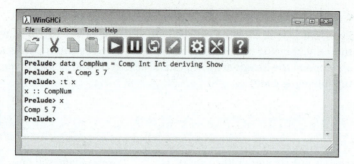

FIGURE 10-2:
Use the deriving keyword to add features to the data type.

Using OR

One of the more interesting aspects of Haskell data types is that you can create a *Sum* data type — a type that contains multiple constructors that essentially define multiple associated types. To create such a type, you separate each data constructor using a bar (|), which is essentially an OR operator. The following code shows how you might create a version of CompNum (shown in the previous section) that provides for complex, purely real, and purely imaginary numbers:

```
data CompNum = Comp Int Int | Real Int | Img Int deriving
    Show
```

When working with a real number, the imaginary part is always 0. Likewise, when working with an imaginary number, the real part is always 0. Consequently, the Real and Img definitions require only one Int as input. Figure 10-3 shows the new version of CompNum in action.

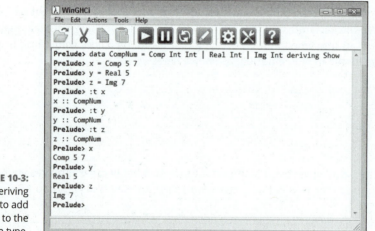

FIGURE 10-3:
Use the deriving keyword to add features to the data type.

As you can see, you define each of the variables using the applicable data constructor. When you check type using :t, you see that they all use the same type constructor: CompNum. However, when you display the individual values, you see the kind of number that the expression contains.

Defining enumerations

The ability to enumerate values is essential as a part of categorizing. Providing distinct values for a particular real-world object's properties is important if you want to better understand the object and show how it relates to other objects in the world. Previous sections explored the use of data constructors with some sort of input, but nothing says that you must provide a value at all. The following code demonstrates how to create an enumeration in Haskell:

```
data Colors = Red | Blue | Green deriving (Show, Eq, Ord)
```

Notice that you provide only a label for the individual constructors that are then separated by an OR operator. As with previous examples, you must use deriving to allow the display of the particular variable's content. Notice, however, that this example also derives from Eq (which tests for equality) and Ord (which tests for inequality). Figure 10-4 shows how this enumeration works.

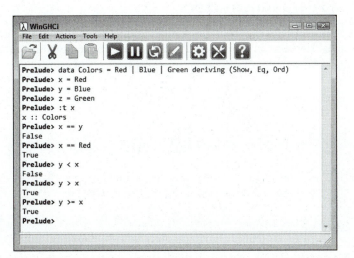

FIGURE 10-4: Enumerations are made of data constructors without inputs.

As usual, the individual variables all use the same data type, which is Colors in this case. You can compare the variable content. For example, x == y is False because they're two different values. Note that you can compare a variable to its data constructor, as in the case of x == Red, which is True. You have access to all of the logical operators in this case, so you could create relatively complex logic based on the truth value of this particular type.

Enumerations also appear using alternative text. Fortunately, Haskell addresses this need as well. This updated code presents the colors in a new way:

```
data Colors = Red | Blue | Green deriving (Eq, Ord)
instance Show Colors where
    show Red = "Fire Engine Red"
    show Blue = "Sky Blue"
    show Green = "Apple Green"
```

The instance keyword defines a specific manner in which instances of this type should perform particular tasks. In this case, it defines the use of Show. Each color appears in turn with the color to associate with it. Notice that you don't define Show in deriving any longer; you use the deriving or instance form, but not both. Assuming that you create three variables as shown in Figure 10-4, (where x = Red, y = Blue, and z = Green), here's the output of this example:

```
x = Fire Engine Red
y = Sky Blue
z = Apple Green
```

Considering type constructors and data constructors

Many data sources rely on records to package data for easy use. A record has individual elements that you use together to describe something. Fortunately, you can create record types in Haskell. Here's an example of such a type:

```
data Name = Employee {
    first :: String,
    middle :: Char,
    last :: String} deriving Show
```

The Name type includes a data constructor for Employee that contains fields named first and last of type String and middle of type Char.

```
newbie = Employee "Sam" 'L' "Wise"
```

Notice that the 'L' must appear in single quotes to make it the Char type, while the other two entries appear in double quotes to make them the String type. Because you've derived Show, you can display the record, as shown in Figure 10-5. Just in case you're wondering, you can also display individual field values, as shown in the figure.

FIGURE 10-5:
Haskell supports record types using special data constructor syntax.

The problem with this construction is that it's rigid, and you may need flexibility. Another way to create records (or any other type, for that matter) is to add the arguments to the type constructor instead, as shown here:

```
data Name f m l = Employee {
    first :: f,
    middle :: m,
    last :: l} deriving Show
```

This form of construction is parameterized, which means that the input comes from the type constructor. The difference is that you can now create the record using a Char or a String for the middle name. Unfortunately, you can also create Employee records that really don't make any sense at all, as shown in Figure 10-6, unless you create a corresponding type signature of Name :: (String String String) -> Employee.

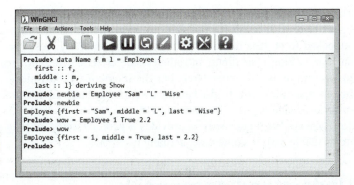

FIGURE 10-6:
Parameterized types are more flexible.

REMEMBER

Haskell supports an incredibly rich set of type structures, and this chapter doesn't do much more than get you started on understanding them. The article at https://wiki.haskell.org/Constructor provides some additional information about type constructors and data constructors, including the use of recursive types.

Composing Types

The following sections talk about composing special types: monoids, monads, and semigroups. What makes these types special is that they have a basis in math, as do most things functional; this particular math, however, is about abstracting away details so that you can see the underlying general rules that govern something and then develop code to satisfy those rules.

REMEMBER

The reason you want to perform the abstraction process is that it helps you create better code with fewer (or possibly no) side effects. Aren't functional languages supposed to be free of side effects, though? Generally, yes, but some activities, such as getting user input, introduces side effects. The math part of functional programming is side-effect free, but the moment you introduce user interaction (as an example), you begin having to perform tasks in a certain order, which introduces side effects. The article at https://wiki.haskell.org/Haskell_IO_for_Imperative_Programmers provides a good overview of why side effects are unavoidable and, in some case, actually necessary.

Understanding monoids

The "Considering the math basis for monoids and semigroups" sidebar may still have you confused. Sometimes an example works best to show how something actually works, instead of all the jargon used to describe it. So, this section begins with a Haskell list, which is a monoid, as it turns out. To prove that it's a monoid, a list has to follow three laws:

>> **Closure:** The result of an operation must always be a part of the set comprising the group that defines the monoid.

>> **Associativity:** The order in which operations on three or more objects occur shouldn't matter. However, the order of the individual elements can matter.

>> **Identity:** There is always an operation that does nothing.

CONSIDERING THE MATH BASIS FOR MONOIDS AND SEMIGROUPS

Monoids and semigroups ultimately belong to abstract algebra and discrete mathematics, as shown at http://www.euclideanspace.com/maths/discrete/index.htm. These are somewhat scary-sounding terms to most people. However, you can view abstractions in a simple way. Say that you look at the picture of three bears on a computer. When asked, the computer will reveal that it's managing millions of pixels — a difficult task. However, when someone asks you the same question, you say you see three bears. In a moment, you have abstracted away the details (millions of pixels with their various properties) and come to a new truth (three bears).

Math abstraction goes even further. In the example of the three bears, a math abstraction would remove the background because it wants to focus on the individual bears (becoming discrete). It might then remove the differences among the animals and eventually remove the animal features of the image so that you end up with an outline showing the essence of the bears — what makes bears in this picture unique — a generalization of those bears. You can then use those bears to identify other bears in other pictures. What the math is doing is helping you generalize the world around you; you really aren't performing arithmetic.

The next level down from math abstraction, as described in this chapter, is the use of groups (see http://www.euclideanspace.com/maths/discrete/groups/index.htm). A *group* is a set of objects that relies on a particular operation to combine pairs of objects within the set. Many of the texts you may read talk about this task using numbers because defining the required rules is easier using numbers. However, you can use any object. Say that you have the set of all letters and the operation of concatenation (essentially letter addition). A word, then, would be the concatenation of individual letters found in the set — the group of all letters.

(continued)

(continued)

The concept of groups always involves like objects found in a set with an associated operation, but beyond this definition, the objects can be of any type, the operation can be of any type, and the result is based on the type of the object and the operation used to combine them. However, groups have specific rules that usually rely on numbers, such as the identity rule, which is an operation that doesn't do anything. For example, adding 0 to a group of numbers doesn't do anything, so using the 0 element with the add operation would satisfy the identity rule. The inverse operation provides what amounts to the negative of the group. For example, in working with the set of all numbers and the add operation, if you combine 1 with –1, you receive 0, the identity element, back.

To create a group that is the concatenation of letters, you need a monoid, as described at http://www.euclideanspace.com/maths/discrete/groups/monoid/index.htm. A *monoid* is like a group except that it doesn't require an inverse operation. There isn't a –a, for example, to go with the letter a. Consequently, you can create words from the set of all letters without having to provide an inverse operation for each word. A *semigroup* is actually a special kind of monoid except that it doesn't include the identity operation, either. In considering the group of all letters, a group that lacks a null character (the identity element) would require a semigroup for expression.

Lists automatically address the first law. If you're working with a list of numbers, performing an operation on that list will result in a numeric output, even if that output is another list. In other words, you can't create a list of numbers, perform an operation on it, and get a Char result. To demonstrate the other two rules, you begin by creating the following three lists:

```
a = [1, 2, 3]
b = [4, 5, 6]
c = [7, 8, 9]
```

In this case, the example uses concatenation (++) to create a single list from the three lists. The associativity law demands that the order in which an operation occurs shouldn't matter, but that the order of the individual elements can matter. The following two lines test both of these criteria:

```
(a ++ b) ++ c == a ++ (b ++ c)
(a ++ b) ++ c == (c ++ b) ++ a
```

The output of the first comparison is True because the order of the concatenation doesn't matter. The output of the second comparison is False because the order of the individual elements does matter.

The third law, the identity law, requires the use of an empty list, which is equivalent to the 0 in the set of all numbers that is often used to explain identity. Consequently, both of these statements are true:

```
a ++ [] == a
[] ++ a == a
```

When performing tasks using some Haskell, you need to use import Data.Monoid. This is the case when working with strings. As shown in Figure 10-7, strings also work just fine as monoids. Note the demonstration of identity using an empty string. In fact, many Haskell collection types work as monoids with a variety of operators, including Sequence, Map, Set, IntMap, and IntSet. Using the custom type examples described earlier in the chapter as a starting point, any collection that you use as a basis for a new type will automatically have the monoid functionality built in. The example at https://www.yesodweb.com/blog/2012/10/generic-monoid shows a more complex Haskell implementation of monoids as a custom type (using a record in this case).

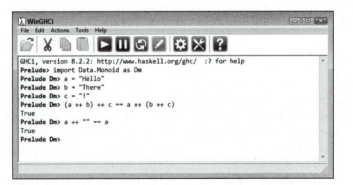

FIGURE 10-7:
Strings can act as
monoids, too.

TIP

After you import Data.Monoid, you also have access to the <> operator to perform append operations. For example, the following line of Haskell code tests the associative law:

```
(a <> b) <> c == a <> (b <> c)
```

REMEMBER

Even though this section has focused on the simple task of appending one object to another, most languages provide an assortment of additional functions to use with monoids, which is what makes monoids particularly useful. For example, Haskell provides the Dual function, which reverses the output of an append operation. The following statement is true because the right expression uses the Dual function:

```
((a <> b) <> c) == getDual ((Dual c <> Dual b) <> Dual a)
```

Even though the right side would seem not to work based on earlier text, the use of the Dual function makes it possible. To make the statement work, you must also call getDual to convert the Dual object to a standard list. You can find more functions of this sort at http://hackage.haskell.org/package/base-4.11.1.0/docs/Data-Monoid.html.

The same rules for collections apply with Python. As shown in Figure 10-8, Python lists behave in the same manner as Haskell lists.

```
■ Anaconda Prompt - python

(base) C:\Users\John>python
Python 3.6.4 |Anaconda, Inc.| (default, Jan 16 2018, 10:22:32) [MSC v.1900 64 bi
t (AMD64)] on win32
Type "help", "copyright", "credits" or "license" for more information.
>>> a = [1, 2, 3]
>>> b = [4, 5, 6]
>>> c = [7, 8, 9]
>>> (a + b) + c == a + (b + c)
True
>>> a + [] == a
True
>>> [] + a == a
True
>>> _
```

FIGURE 10-8: Python collections can also act as monoids.

TIP

In contrast to Haskell, Python doesn't have a built-in monoid class that you can use as a basis for creating your own type with monoid support. However, you can see plenty of Python monoid implementations online. The explanation at https://github.com/justanr/pynads/blob/master/pynads/abc/monoid.py describes how you can implement the Haskell functionality as part of Python. The implementation at https://gist.github.com/zeeshanlakhani/1284589 is shorter and probably easier to use, plus it comes with examples of how to use the class in your own code.

Considering the use of Nothing, Maybe, and Just

Haskell doesn't actually have a universal sort of null value. It does have Nothing, but to use Nothing, the underlying type must support it. In addition, Nothing is actually something, so it's not actually null (which truly is nothing). If you assign Nothing to a variable and then print the variable onscreen, Haskell tells you that its value is Nothing. In short, Nothing is a special kind of value that tells you that the data is missing, without actually assigning null to the variable. Using this approach has significant advantages, not the least of which is fewer application crashes and less potential for a missing value to create security holes.

You normally don't assign Nothing to a variable directly. Rather, you create a function or other expression that makes the assignment. The following example shows a simple function that simply adds two numbers. However, the numbers must be positive integers greater than 0:

```
doAdd::Int -> Int -> Maybe Int
doAdd _ 0 = Nothing
doAdd 0 _ = Nothing
doAdd x y = Just (x + y)
```

Notice that the type signature has Maybe Int as the output. This means that the output could be an Int or Nothing. Before you can use this example, you need to load some support for it:

```
import Data.Maybe as Dm
```

To test this how Maybe works, you can try various versions of the function call:

```
doAdd 5 0
doAdd 0 6
doAdd 5 6
```

The first two result in an output of Nothing. However, the third results in an output of Just 11. Of course, now you have a problem, because you can't use the output of Just 11 as numeric input to something else. To overcome this problem, you can make a call to fromMaybe 0 (doAdd 5 6). The output will now appear as 11. Likewise, when the output is Nothing, you see a value of 0, as shown in Figure 10-9. The first value to fromMaybe, 0, tells what to output when the output of the function call is Nothing. Consequently, if you want to avoid the whole Nothing issue with the next call, you can instead provide a value of 1.

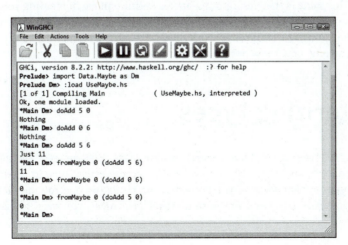

FIGURE 10-9: Haskell enables you to process data in unique ways with little code.

TIP

As you might guess, Python doesn't come with `Maybe` and `Just` installed. However, you can add this functionality or rely on code that others have created. The article at `http://blog.senko.net/maybe-monad-in-python` describes this process and provides a link to a `Maybe` implementation that you can use with Python. The PyMonad library found at `https://pypi.org/project/PyMonad/` also includes all of the required features and is easy to use.

Understanding semigroups

The "Understanding monoids" section, earlier in this chapter, discusses three rules that monoids must follow. Semigroups are like monoids except that they have no identity requirement. Semigroups actually represent a final level of abstraction, as discussed in the earlier sidebar, "Considering the math basis for monoids and semigroups". At this final level, things are as simple and flexible as possible. Of course, sometimes you really do need to handle a situation in which something is `Nothing`, and the identity rule aids in dealing with this issue. People have differing opinions over the need for and usefulness of semigroups, as shown in the discussion at `https://stackoverflow.com/questions/40688352/why-prefer-monoids-over-semigroups-in-haskell-why-do-we-need-mempty`. However, a good rule of thumb is to use the simplest abstraction when possible, which would be semigroups whenever possible. To work with semigroups, you must execute `import Data.Semigroup`.

REMEMBER

You may wonder why you would use a semigroup when a monoid seems so much more capable. An example of an object that must use a semigroup is a bounding box. A bounding box can't be empty; it must take up some space or it doesn't exist and therefore the accompanying object has no purpose. Another example of when to use a semigroup is `Data.List.NonEmpty` (`http://hackage.haskell.org/package/base-4.11.1.0/docs/Data-List-NonEmpty.html`), which is a list that must always have at least one entry. Using a monoid in this case wouldn't work. The point is that semigroups have a definite place in creating robust code, and in some cases, you actually open your code to error conditions by not using them. Fortunately, semigroups work much the same as monoids, so if you know how to use one, you know how to use the other.

Parameterizing Types

The "Considering type constructors and data constructors" section, earlier in this chapter, shows you one example of a parameterized type in the form of the `Name` type. In that section, you consider two kinds of constructions for the `Name` type that essentially end in the same result. However, you need to use parameterized types at the right time. Parameterized types work best when the type acts as a sort of box that could hold any sort of value. The Name type is pretty specific, so it's not the best type to parameterize because it really can't accept just any kind of input.

A better example for parameterizing types would be to create a custom tuple that accepts three inputs and provides the means to access each member using a special function. It would be sort of an extension of the fst and snd functions provided by the default tuple. In addition, when creating a type of this sort, you want to provide some sort of conversion feature to a default. Here is the code used for this example:

```
data Triple a b c = Triple (a, b, c) deriving Show

fstT (Triple (a, b, c)) = show a
sndT (Triple (a, b, c)) = show b
thdT (Triple (a, b, c)) = show c

cvtToTuple (Triple (a, b, c)) = (a, b, c)
```

In this case, the type uses parameters to create a new value: a, b, and c represent elements of any type. Consequently, this example starts with a real tuple, but of a special kind, Triple. When you display the value using show, the output looks like any other custom type.

The special functions enable you to access specific elements of the Triple. To avoid name confusion, the example uses a similar, but different, naming strategy of fstT, sndT, and thdT. Theoretically, you could use wildcard characters for each of the nonessential inputs, but good reason exists to do so in this case.

Finally, cvtToTuple enables you to change a Triple back into a tuple with three elements. The converted tuple has all the same functionality as a tuple that you create any other way. The following test code lets you check the operation of the type and associated functions:

```
x = Triple("Hello", 1, True)

show(x)
fstT(x)
sndT(x)
thdT(x)
show(cvtToTuple(x)))
```

The outputs demonstrate that the type works as expected:

```
Triple ("Hello",1,True)
"Hello"
1
True
("Hello",1,True)
```

Unfortunately, there isn't a Python equivalent of this code. You can mimic it, but you must create a custom solution. The material at https://ioam.github.io/param/Reference_Manual/param.html#parameterized-module and https://stackoverflow.com/questions/46382170/how-can-i-create-my-own-parameterized-type-in-python-like-optionalt is helpful, but this is one time when you may want to rely on Haskell if this sort of task is critical for your particular application and you don't want to create a custom solution.

Dealing with Missing Data

In a perfect world, all data acquisition would result in complete records with nothing missing and nothing wrong. However, in the real world, datasets often contain a lot of missing data, and you're often left wondering just how to address the issue so that your analysis is correct, your application doesn't crash, and no one from the outside can corrupt your setup using something like a virus. The following sections don't handle every possible missing-data issue, but they give you an overview of what can go wrong as well as offer possible fixes for it.

Handling nulls

Different languages use different terms for the absence of a value. Python uses the term None and Haskell uses the term Nothing. In both cases, the value indicates an absence of an anticipated value. Often, the reasons for the missing data aren't evident. The issue is that the data is missing, which means that it's not available for use in analysis or other purposes. In some languages, the missing value can cause crashes or open a doorway to viruses (see the upcoming "Null values, the billion-dollar mistake" sidebar for more information).

When working with Haskell, you must provide a check for Nothing values, as described in the "Considering the use of Maybe and Just" section, earlier in this chapter. The goal is to ensure that the checks in place now that a good reason for unchecked null values no longer exist. Of course, you must still write your code proactively to handle the Nothing case (helped by the Haskell runtime that ensures that functions receive proper values). The point is that Haskell doesn't have an independent type that you can call upon as Nothing; the Nothing type is associated with each data type that requires it, which makes locating and handling null values easier.

Python does include an individual null type called None, and you can assign it to a variable. However, note that None is still an object in Python, although it's not in other languages. The variable still has an object assigned to it: the None object. Because None is an object, you can check for it using is. In addition, because of the nature of None, it tends to cause fewer crashes and leave fewer doors open to nefarious individuals. Here is an example of using None:

```
x = None
if x is None:
    print("x is missing")
```

REMEMBER

The output of this example is x is missing, as you might expect. You should also note that Python lacks the concept of pointers, which is a huge cause of null values in other languages. Someone will likely point out that you can also check for None using x == None. This is a bad idea because you can override the == (equality) operator but you can't override is, which means that using is provides a consistent behavior. The discussion at https://stackoverflow.com/questions/3289601/null-object-in-python provides all the details about the differences between == and is and why you should always use is.

NULL VALUES, THE BILLION-DOLLAR MISTAKE

Null values cause all sorts of havoc in modern-day applications. However, they were actually started as a means of allowing applications to run faster on notoriously slow equipment. The checks required to ensure that null values didn't exist took time and could cause applications to run absurdly slowly. Like most fixes for problems with speed, this one comes with a high cost that continues to create problems such as opening doors to viruses and causing a host of tough-to-locate data errors. The presentation at https://www.infoq.com/presentations/Null-References-The-Billion-Dollar-Mistake-Tony-Hoare calls null references a billion-dollar mistake. This presentation that helps developers understand the history, and therefore the reasoning, behind null values that are now a plague in modern application development.

Performing data replacement

Missing and incorrect data present problems. Before you can do anything at all, you must verify the dataset you use. Creating types (using the techniques found in earlier sections in the chapter) that automatically verify their own data is a good start. For example, you can create a type for bank balances that ensure that the balance is never negative (unless you want to allow an overdraft). However, even with the best type construction available, a dataset may contain unusable data entries or some entries that don't contain data at all. Consequently, you must per-form verification of such issues as missing data and data that appears out of range.

After you find missing or incorrect data, you consider the ramifications of the error. In most cases, you have the following three options:

>> Ignore the issue

>> Correct the entry

>> Delete the entry and associated elements

Ignoring the issue might cause the application to fail and will most certainly pro-duce inaccurate results when the entry is critical for analysis. However, most datasets contain superfluous entries — those that you can ignore unless you require the amplifying information they provide.

Correcting the entry is time consuming in most cases because you must now define a method of correction. Because you don't know what caused the data error in the first place, or the original data value, any correction you make will be flawed to some extent. Some people use statistical measures (as described in the next section) to make a correction that neither adds to nor detracts from the overall statistical picture of the entries taken together. Unfortunately, even this approach is flawed because the entry may have represented an important departure from the norm.

Deleting the entry is fast and fixes the problem in a way that's unlikely to cause the application to crash. However, deleting the entry comes with the problem of affecting any analysis you perform. In addition, deleting an entire row (case) from a dataset means losing not only the corrected entry (the particular feature) but also all the other entries in that row. Consequently, deletion of a row can cause noticeable data damage in some cases.

Considering statistical measures

A *statistical measure* is one that relies on math to create some sort of overall or average entry to use in place of a missing or incorrect entry. Depending on the data in question and the manner in which you create types to support your application, you may be able to rely on statistical measures to fix at least some problems in your dataset.

REMEMBER

Statistical measures generally see use for only numeric data. For example, guessing about the content of a string field would be impossible. If the analysis you perform on the string field involves a numeric measure such as length or frequency of specific letters, you might use statistical measures to create a greeked text (essentially nonsense text) replacement (see http://www.webdesignerstoolkit.com/copy.php for details), but you can't create the actual original text.

Some statistical corrections for missing or inaccurate data see more use than others do. In fact, you can narrow the list of commonly used statistical measures down to these:

» **Average (or mean):** A calculation that involves adding all the values in a column together and dividing by the number of items in that column. The result is a number that is the average of all the numbers in the column. This is the measure that is least likely to affect your analysis.

» **Median:** The middle value of a series of numbers. This value is not necessarily an average but is simply a middle value. For example, in the series 1, 2, 4, 4, 5, the value 4 is the median because it appears in the middle of the set. The average (or mean) would be 3.2 instead. This is the measure that is most likely to represent the middle value and generally affects the analysis only slightly.

» **Most common (mode):** The number that appears most often in a series, even if the value is at either end of the scale. For example, in the series 1, 1, 1, 2, 4, 5, 6, the mode is 1, the average is 2.8, and the median is 2. This is the measure that reflects the value that has the highest probability of being correct, even if it affects your analysis significantly.

As you can see, using the right statistical measure is important. Of course, there are many other statistical measures, and you may find that one of them fits your data better. A technique that you can use to ensure that especially critical values are the most accurate possible is to plot the data to see what shape it creates and then use a statistical measure based on shape.

Creating and Using Type Classes

Haskell has plenty of type classes. In fact, you use them several times in this chapter. The most common type classes include Eq, Ord, Show, Read, Enum, Bounded, Num, Integral, and Floating. The name *type class* confuses a great many people — especially those with an Object-Oriented Programming (OOP) background.

In addition, some people confuse type classes and types such as Int, Float, Double, Bool, and Char. Perhaps the best way to view a type class is as a kind of interface in which you describe what to do but not how to do it. You can't use a type class directly; rather, you derive from it. The following example shows how to use a type class named Equal:

```
class Equal a where (##) :: a -> a -> Bool

data MyNum = I Int deriving Show
instance Equal MyNum where
   (I i1) ## (I i2) = i1 == i2
```

In this case, Equal defines the ## operator, which Haskell doesn't actually use. Equal accepts two values of any type, but of the same types (as defined by a) and outputs a Bool. However, other than these facts, Equal has no implementation.

MyNum, a type, defines I as accepting a single Int value. It derives from the common type class, Show, and then implements an instance of Equal. When creating your own type class, you must create an implementation for it in any type that will use it. In this case, Equal simply checks the equality of two variables of type MyNum. You can use the following code to test the result:

```
x = I 5
y = I 5
z = I 6

x ## y
x ## z
```

In the first case, the comparison between x and y, you get True as the output. In the second case, the comparison of x and z, you get False as the output. Type classes provide an effective means of creating common methods of extending basic type functionality. Of course, the implementation of the type class depends on the needs of the deriving type.

4

Interacting in Various Ways

Chapter **11**

Performing Basic I/O

To be useful, most applications must perform some level of Input/Output (I/O). Interaction with the world outside the application enables the application to receive data (input) and provide the results of any operations performed on that data (output). Without this interaction, the application is self-contained, and although it could conceivably perform work, that work would be useless. Any language that you use to create a useful application must support I/O. However, I/O would seem to be counter to the functional programming paradigm because most languages implement it as a procedure — a process. But functional languages implement I/O differently from other languages; they use it as a pure function. The goal is to implement I/O without side effects, not to keep I/O from occurring. The first part of this chapter discusses how I/O works in the functional programming paradigm.

After you know how the I/O process works, you need some means of managing the data. This chapter begins by looking at the first kind of I/O that most applications perform, data input, and then reviews data output. You discover how the functional programming paradigm makes I/O work without the usual side effects. This first part also discusses some differences in device interactions.

TIP

Jupyter Notebook offers magic functions that make working with I/O easier. This chapter also looks at the features provided by magic functions when you're working Python. Because Jupyter Notebook provides support for a long list of languages, you may eventually be able to use magic functions with Haskell as well.

The final part of this chapter puts together everything you've discovered about I/O in the functional programming paradigm. You see how Haskell and Python handle the task in both pure and impure ways. Performing I/O and programming in a functional way aren't mutually exclusive, and no one is breaking the rules to make it happen. However, each language has a slightly different approach to the issue, so a good understanding of each approach is important.

Understanding the Essentials of I/O

Previous chapters discuss the essentials of the functional programming paradigm. Some of these issues are mechanical, such as the immutability of data. In fact, some would argue that these issues aren't important — that only the use of pure functions is important. The various coding examples and explanations in those previous chapters tend to argue otherwise, but for now, consider only the need to perform I/O using pure functions that produce no side effects. In some respects, that really isn't possible. (Some people say it is, but the proof is often lacking.) The following sections discuss I/O from a functional perspective and help you understand the various sides of the argument over whether performing I/O using pure functions is possible.

Understanding I/O side effects

An essential argument that many people make regarding I/O side effects is actually quite straightforward. When you create a function in a functional language and apply specific inputs, you receive the same answer every time, as long as those inputs remain the same. For example, if you calculate the square root of 4 and then make the same call 99 more times, you receive the answer 2 every time. In fact, a language optimizer would do well to simply cache the result, rather than perform the calculation, to save time.

However, if you make a call to query the user for input, you receive a certain result. Making the same call, with the same query, 99 more times may not always produce the same result. For example, if you pose the question "What is your name?" the response will differ according to user. In fact, the same user could answer differently by providing a full name one time and only a first name another. The fact that the function call potentially produces a different result with each call is a side effect. Even though the developer meant for the side effect to occur, from the definitions of the functional programming paradigm in past chapters, I/O produces a side effect in this case.

The situation becomes worse when you consider output. For example, when a function makes a query to the user by outputting text to the console, it has changed the state of the system. The state is permanently changed because returning the system to its previous state is not possible. Even removing the characters would mean making a subsequent change.

Unfortunately, because I/O is a real-world event, you can't depend on the occurrence of the activity that you specify. When you calculate the square root of 4, you always receive 2 because you can perform the task as a pure function. However, when you ask the user for a name, you can't be sure that the user will supply a name; the user might simply press Enter and present you with nothing. Because I/O is a real-world event with real-world consequences, even functional languages must supply some means of dealing with the unexpected, which may mean exceptions — yet another side effect.

Many languages also support performing I/O separately from the main application thread so that the application can remain responsive. The act of creating a thread changes the system state. Again, creating a thread is another sort of side effect that you must consider when performing I/O. You need to deal with issues such as the system's incapability to support another thread or knowing whether any other problems arose with the thread. The application may need to allow inter-thread communication, as well as communication designed to ascertain thread status, all of which requires changing application state.

REMEMBER

This section could continue detailing potential side effects because myriad side effects are caused by I/O, even successful I/O. Functional languages make a clear distinction between pure functions used to perform calculations and other internal tasks and I/O used to affect the outside world. The use of I/O in any application can potentially cause these problems:

>> **No actual divide:** Any function can perform I/O when needed. So the theoretical divide between pure functions and I/O may not be as solid as you think.

>> **Monolithic:** Because I/O occurs in a sequence (you can't obtain the next answer from a user before you obtain the current answer), the resulting code is monolithic and tends to break easily. In addition, you can't cache the result of an I/O; the application must perform the call each time, which means that optimizing I/O isn't easy.

>> **Testing:** All sorts of issues affect I/O. For example, an environmental condition (such as lightning) that exists now and causes an I/O to fail may not exist five minutes from now.

>> **Scaling:** Because I/O changes system state and interacts with the real world, the associated code must continue executing in the same environment. If the system load suddenly changes, the code will slow as well because scaling the code to use other resources isn't possible.

TIP

The one way you have to overcome these problems in a functional environment is to ensure that all the functions that perform I/O remain separate from those that perform calculations. Yes, the language you use may allow the mixing and matching of I/O and calculations, but the only true way around many of these problems is to enforce policies that ensure that the tasks remain separate.

Using monads for I/O

The "Understanding monoids" section of Chapter 10 discusses monads and their use, including strings. Interestingly enough, the IO class in Haskell, which provides all the I/O functionality, is a kind of monad, as described at https://hackage.haskell.org/package/base-4.11.1.0/docs/System-IO.html. Of course, this sounds rather odd, but it's a fact. Given what you know about monads, you need to wonder what the two objects are and what the operator is. In looking down the list of functions for the IO class, you discover that IO is the operator. The two objects are a handle and the associated data.

REMEMBER

A *handle* is a method for accessing a device. Some handles, such as stderr, stdin, and stdout, are standard for the system, and you don't need to do anything special to use them. For both Python and Haskell, these standard handles point to the keyboard for stdin and the display (console) for stdout and stderr. Other handles are unique to the destination, such as a file on the local drive. You must first acquire the handle (including providing a description of how you plan to use it) and then add it to any call you make.

Interacting with the user

The concept of using a monad for I/O has some ramifications that actually make Haskell I/O easier to understand, despite its being essentially the same as any other I/O you might have used. When performing input using getLine, what you really do is combine the stdin handle with the data the user provides using the IO operator. Yes, it's the same thing you do with the Python input method, but the underlying explanation for the action is different in the two cases; when working with Python, you're viewing the task as a procedure, not as a function. To see how this works, type **:t getLine** and press Enter. You see that the type of getLine (a function) is IO String. Likewise, type **:t putStrLn** and press Enter, and you see that the type of putStrLn is String -> IO (). However, when you use the following code:

```
putStrLn "What is your name?"
name <- getLine
putStrLn $ "Hello " ++ name
```

you obtain the same result as you might expect from any programming language. Only the manner in which you review the action differs, not the actual result of the action, as shown in Figure 11-1.

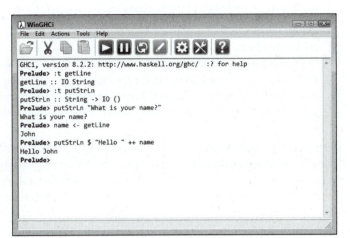

FIGURE 11-1:
Interacting with the user implies using monads with an operator of IO.

Notice that you must use the apply operator ($) to the second putStrLn call because you need to apply the result of the monad "Hello " ++ name (with ++ as the operator) to putStrLn. Otherwise, Haskell will complain that it was expecting a [char]. You could also use putStrLn ("Hello " ++ name) in place of the apply operator.

Working with devices

Always remember that humans interact with devices — not code, not applications, and not actually with data. You could probably come up with a lot of different ways to view devices, but the following list provides a quick overview of the essential device types:

» **Host:** The host device is the system on which the application runs. Most languages support standard inputs and outputs for the host device that don't require any special handling.

» **Input:** Anything external to the host can provide input. In this case, external to the host means anything outside the localized processing environment, including hard drives housed within the same physical structure as the motherboard that supports the host. However, inputs can come from anywhere, including devices such as external cameras from a security system.

>> **Output:** An output device can be anything, including a hard drive within the same physical case as the host. However, outputs also include physical devices outside the host case. For example, sending a specific value to a robot may create a thousand widgets. The I/O has a distinct effect on the outside world outside the realm of the host device.

>> **Cloud:** A *cloud device* is one that doesn't necessarily have physicality. The device could be anywhere. Even if the device must have a physical presence (such as a hard drive owned by a host organization), you may not know where the device is located and likely don't even care. People are using more and more cloud devices for everything from data storage to website hosting, so you're almost certain to deal with some sort of cloud environment.

All the I/O that you perform with a programming language implies access to a device, even when working with a host device. For example, when working with Haskell, the hPutStrLn and putStrLn lines of code that follow are identical in effect (note that you must import System.IO before you can perform this task):

```
import System.IO as IO
hPutStrLn stdout "Hello there!"
putStrLn "Hello there!"
```

The inclusion of stdout in the first call to hPutStrLn simply repeats what putStrLn does without an explicit handle. However, in both cases, you do need a handle to a device, which is the host in this case. Because the handle is standard, you don't need to obtain one.

Getting a handle for a local device is relatively easy. The following code shows a three-step process for writing to a file:

```
import System.IO as IO
handle <- openFile "MyData.txt" WriteMode
hPutStrLn handle "This is some test data."
hClose handle
```

When calling openFile, you again use the IO operator. This time, the two objects are the file path and the I/O mode. The output, when accessing a file successfully, is the I/O handle. Haskell doesn't use the term *file handle* as other languages do because the handle need not necessarily point to a file. As always, you can use :t openFile to see the definition for this function. When you don't supply a destination directory, GHCi resorts to using whatever directory you have assigned for loading files. Here is the code used to read the content from the file:

```
import System.IO as IO
handle <- openFile "MyData.txt" ReadMode
myData <- hGetLine handle
hClose handle
putStrLn myData
```

TECHNICAL STUFF

This chapter doesn't fully explore everything you can do with various I/O methodologies in Haskell. For example, you can avoid getting a handle to read and write files by using the readFile, writeFile, and appendFile functions. These three functions actually reduce the three-step process into a single step, but the same steps occur in the background. Haskell does support the full range of device-oriented functions for I/O found in other languages.

Manipulating I/O Data

This chapter doesn't discuss all the ins and outs of data manipulation for I/O purposes, but it does give you a quick overview of some issues. One of the more important issues is that both Haskell and Python tend to deal with string or character output, not other data types. Consequently, you must convert all data you want to output to a string or a character. Likewise, when you read the data from the source, you must convert it back to its original form. A call, such as appendFile "MyData.txt" 2, simply won't work. The need to work with a specific data type contrasts to other operations you can perform with functional languages, which often assume acceptance of any data type. When creating functions to output data, you need to be aware of the conversion requirement because sometimes the error messages provided by the various functional languages are less than clear as to the cause of the problem.

Another issue is the actual method used to communicate with the outside world. For example, when working with files, you need to consider *character encoding* (the physical representation of the characters within the file, such as the number of bits used for each character). Both Haskell and Python support a broad range of encoding types, including the various Unicode Transformation Format (UTF) standards described at https://www.w3.org/People/danield/unic/unitra.htm. When working with text, you also need to consider issues such as the method used to indicate the end of the line. Some systems use both a carriage return and line feed; others don't.

Devices may also require the use of special commands or headers to alert the device to the need to communicate and establish the communication methods. Neither Haskell nor Python has these sorts of needs built into the language, so you must either create your own solution or rely on a third-party library. Likewise, when working with the cloud, you often must provide the data in a specific format and include headers to describe how to communicate and with which service to communicate (among other things).

REMEMBER

The reason for considering all these issues before you try to communicate is that a large number of online help messages deal with these sorts of issues. The language works as intended in producing output or attempting to receive input, but the communication doesn't work because of the lack of communication *protocol* (a set of mutually acceptable rules). Unfortunately, the rules are so diverse and some so arcane as to defy any sort of explanation in a single book. Make sure to keep in mind that communication is often a lot more than simply sending or receiving data, even in a functional language in which some things seem to happen magically.

Using the Jupyter Notebook Magic Functions

Python can make your I/O experience easier when you work with specific tools, which is the point of this section. Notebook and its counterpart, IPython, provide you with some special functionality in the form of magic functions. It's kind of amazing to think that these applications offer you magic, but that's precisely what you get with the magic functions. The magic is in the output. For example, instead of displaying graphic output in a separate window, you can choose to display it within the cell, as if by magic (because the cells appear to hold only text). Or you can use magic to check the performance of your application, and do so without all the usual added code that such performance checks require.

A *magic function* begins with either a percent sign (%) or double percent sign (%%). Those with a % sign work within the environment, and those with a %% sign work at the cell level. For example, if you want to obtain a list of magic functions, type **%lsmagic** and then press Enter in IPython (or run the command in Notebook) to see them, as shown in Figure 11-2. (Note that IPython uses the same input, In, and output, Out, prompts that Notebook uses.)

REMEMBER

Not every magic function works with IPython. For example, the %autosave function has no purpose in IPython because IPython doesn't automatically save anything.

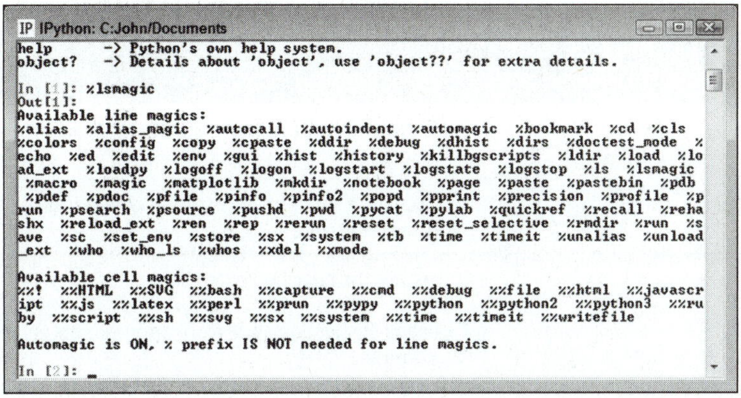

FIGURE 11-2:
The %lsmagic function displays a list of magic functions for you.

Table 11-1 lists a few of the most common magic functions and their purpose. To obtain a full listing, type **%quickref** and press Enter in Notebook (or the IPython console) or check out the full listing at https://damontallen.github.io/IPython-quick-ref-sheets/.

TABLE 11-1

Common Notebook and IPython Magic Functions

Magic Function	Type Alone Provides Status?	Description
%alias	Yes	Assigns or displays an alias for a system command.
%autocall	Yes	Enables you to call functions without including the parentheses. The settings are Off, Smart (default), and Full. The Smart setting applies the parentheses only if you include an argument with the call.
%automagic	Yes	Enables you to call the line magic functions without including the percent (%) sign. The settings are False (default) and True.
%autosave	Yes	Displays or modifies the intervals between automatic Notebook saves. The default setting is every 120 seconds.
%cd	Yes	Changes directory to a new storage location. You can also use this command to move through the directory history or to change directories to a bookmark.
%cls	No	Clears the screen.
%colors	No	Specifies the colors used to display text associated with prompts, the information system, and exception handlers. You can choose between NoColor (black and white), Linux (default), and LightBG.
%config	Yes	Enables you to configure IPython.
%dhist	Yes	Displays a list of directories visited during the current session.

(continued)

TABLE 11-1 *(continued)*

Magic Function	Type Alone Provides Status?	Description
%file	No	Outputs the name of the file that contains the source code for the object.
%hist	Yes	Displays a list of magic function commands issued during the current session.
%install_ext	No	Installs the specified extension.
%load	No	Loads application code from another source, such as an online example.
%load_ext	No	Loads a Python extension using its module name.
%lsmagic	Yes	Displays a list of the currently available magic functions.
%magic	Yes	Displays a help screen showing information about the magic functions.
%matplotlib	Yes	Sets the back-end processor used for plots. Using the inline value displays the plot within the cell for an IPython Notebook file. The possible values are: 'gtk', 'gtk3', 'inline', 'nbagg', 'osx', 'qt', 'qt4', 'qt5', 'tk', and 'wx'.
%paste	No	Pastes the content of the Clipboard into the IPython environment.
%pdef	No	Shows how to call the object (assuming that the object is callable).
%pdoc	No	Displays the docstring for an object.
%pinfo	No	Displays detailed information about the object (often more than provided by help alone).
%pinfo2	No	Displays extra detailed information about the object (when available).
%reload_ext	No	Reloads a previously installed extension.
%source	No	Displays the source code for the object (assuming that the source is available).
%timeit	No	Calculates the best performance time for an instruction.
%%timeit	No	Calculates the best performance time for all the instructions in a cell, apart from the one placed on the same cell line as the cell magic (which could therefore be an initialization instruction).
%unalias	No	Removes a previously created alias from the list.
%unload_ext	No	Unloads the specified extension.
%%writefile	No	Writes the contents of a cell to the specified file.

Receiving and Sending I/O with Haskell

Now that you have a better idea of how I/O in the functional realm works, you can find out a few additional tricks to use to make I/O easier. The following sections deal specifically with Haskell because the I/O provided with Python follows the more traditional procedural approach (except in the use of things like lambda functions, which already appear in previous chapters).

Using monad sequencing

Monad sequencing helps you create better-looking code by enabling you to combine functions into a procedure-like entity. The goal is to create an environment in which you can combine functions in a manner that makes sense, yet doesn't necessarily break the functional programming paradigm rules. Haskell supports two kinds of monad sequencing: without value passing and with value passing. Here is an example of monad sequencing without value passing:

```
name <- putStr "Enter your name: " >> getLine
putStrLn $ "Hello " ++ name
```

In this case, the code creates a prompt, displays it onscreen, obtains input from the user, and places that input into name. Notice the monad sequencing operator (>>) between the two functions. The assignment operator works only with output values, so name contains only the result of the call to getLine. The second line demonstrates this fact by showing the content of name.

You can also create monad sequencing that includes value passing. In this case, the direction of travel is from left to right. The following code shows a function that calls getLine and then passes the result of that call to putStrLn.

```
echo = getLine >>= putStrLn
```

To use this function, type **echo** and press Enter. Anything you type as input echoes as output. Figure 11-3 shows the results of these calls.

Employing monad functions

Because Haskell views I/O as a kind of monad, you also gain access to all the monad functions found at http://hackage.haskell.org/package/base-4.11.1.0/docs/Control-Monad.html. Most of these functions don't appear particularly useful until you start using them together. For example, say that you need to replicate a particular string a number of times. You could use the following code to do it:

```
sequence_ (replicate 10 (putStrLn "Hello"))
```

The call to `sequence_` (with an underscore) causes Haskell to evaluate the sequence of monadic actions from left to right and to discard the result. The `replicate` function performs a task repetitively a set number of times. Finally, `putStrLn` outputs a string to `stdout`. Put it all together and you see the result shown in Figure 11-4.

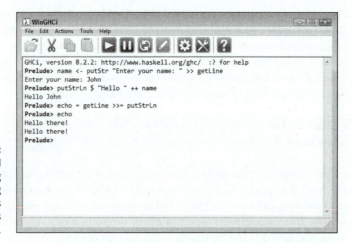

FIGURE 11-3:
Monad sequencing makes combining monad functions in specific ways easier.

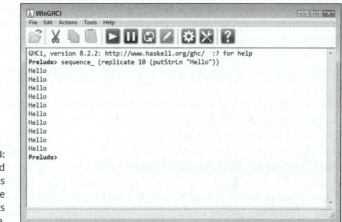

FIGURE 11-4:
Use monad functions to achieve specific results using little code.

Chapter **12**

Handling the Command Line

Working at the command line may seem mildly old fashioned in a world of GUI applications that can perform amazing tricks. However, most developers and administrators know differently. Many of the tools in use today still rely on the command line because it provides a relatively simple, straightforward, and efficient method of interacting with an application. Of course, working at the command line has downsides, too. The most easily understood price of using the command line pertains to ease of use. Anyone who has used the command line extensively knows that it's all too easy to forget command line switches, data inputs, and other required information that a GUI would normally supply as part of a menu entry or form. This chapter begins by discussing methods to make the command line a bit easier to work with.

From the user perspective, remembering arcane command-line syntax is one of the negatives of using the command line. From the developer perspective, finding effective ways to separate the various bits of input and turn them into useful application arguments can sometimes be harder still. The problem for the developer is one of creating an effective interface that provides great flexibility and is forgiving of errant user input (whenever possible). The next part of this chapter talks about using libraries to make working with the command line easier.

Getting Input from the Command Line

When users interact with an application that you create at the command line, they the command line to provide a flexible, simple interface with a certain amount of assistance and robust error detection. Unfortunately, these expectations can be hard to meet, especially that of robust error detection. Trying to create robust command-line error detection can help you better understand the issues faced by people who write compilers because you suddenly face the vagaries of turning text into useful tokens. The following sections help you get started at the command line with a focus on achieving the user goals for application use.

Automating the command line

Even though you see lots of online tutorials that demonstrate utility-type applications used manually, many people simply don't have time or the inclination to type everything manually every time they need a particular application. One of the best features of command-line utilities is that you can automate them in various ways, such as by using batch processing. To automate a command-line utility, you must provide it with a complete set of commands accessible with switches.

The most common switches in use today begin with a slash (/), dash (-), or double dash (--). For example, typing `MyApp -h` could display a help screen for your application. In many cases, the command-line switch is followed by data required to execute the command. The data can be optional or required. For example, `MyApp -h Topic` could display specific help about `Topic`, rather than more generalized help.

Considering the use of prompts

Application developers often feel that adding prompts to the application makes it friendlier. In some respects, adding prompts to ask the user for additional information is better than providing an error message or an error output. However, the use of prompts can also interfere with automation because a batch process won't know what to do with a prompt. Consequently, you must consider the balance between user friendliness and the need to automate when creating a command-line utility. Most people use one of these options when designing their application:

» Avoid using prompts or error messages at all and always provide an error code that is testable in a batch process.

» Use a combination of error messages and error codes to convey the need for additional information without resorting to prompts.

>> Provide a special command-line switch to turn prompts on or off and then rely on one of the first two options in this list when the prompts are off.

>> Employ timed prompts that give the user a specific timeframe in which to respond to queries. A command-line switch can set the interval for displaying the prompt. The application then relies on one of the first two options in this list when the response time expires.

>> Try to obtain the required information using a prompt first, and then rely on a combination of an error message and error code when the user fails to provide the required information on request.

>> Use prompts only, and never provide an error output that could cause potential environmental issues. On failure, the task simply remains undone.

REMEMBER

The choice you make depends on the task your utility performs and on what the user expects from it. For example, a utility that displays the time without doing much else might use the last item on the list without a problem because displaying the time is hardly consequential in most cases. On the other hand, if your utility is performing required analysis of input before the next utility uses the information to configure a set of robotic workers, the first or second option in the list might be better.

Using the command line effectively

A command line utility will interact with the user in a manner that contrasts with a GUI application of the same sort. When working with a GUI, the user has visual aids to understand the relationships among commands. Even if the required command exists several layers down in the menu structure or on a pop-up form, its relationship to other commands is visual. For example, to open a file, you may use the File⇨Open command in a GUI, which requires two mouse clicks, one for each menu level. The speed obtained from using a command-line utility stems partly from not having to deal with a visual interface, thereby letting you access any command at any time without having to delve into the interface at all. Instead of using a File⇨Open command, you may simply specify the filename on the command line, such as `MyApp MyFile`. In addition, command-line utilities allow adding all the commands you want to execute as part of a single command line, making command-line utilities incredibly efficient. For example, say that you want to print the file after you open it. Using a GUI, you might need four mouse clicks: File⇨Open, followed by File⇨Print. A command-line utility needs just one command, `MyApp /p MyFile`, where `/p` is the print switch. Consequently, you must design your command line with the need for brevity and efficiency in mind.

Because users have bad memories, you must provide help with your command-line utility, and convention dictates using the h command-line switch for this purpose. Of course, you precede the h with whatever special symbol you use to designate a command, such as /h, -h, or --h. In addition, most developers allow you to use the question mark (?) to provide access to general help.

TIP

A problem with the help provided with most command-line utilities is that complex utilities often try to answer every question by using a single help screen that goes on for several pages. In some cases, the help screen is so large that it actually scrolls right off the screen buffer, so the developer often tries to solve the problem by adding paging to the help screen. A better option is to provide a general page of help topics and then augment help using individual, short screens for each topic.

Accessing the Command Line in Haskell

The operating system makes certain kinds of information available to applications, such as the command line and environment variable, no matter which language you use to create the applications. Of course, the language must also make access to the information available, but no language is likely to hide the required access because hackers would figure out how to access it anyway. However, it's not always best to use the native information directly. The following sections help you decide how to provide access to command-line arguments in your Haskell application.

Using the Haskell environment directly

Haskell provides access to the operating system environment, including the command-line arguments, in a number of ways. Even though you can find a number of detailed tutorials online, such as the one found at https://wiki.haskell.org/Tutorials/Programming_Haskell/Argument_handling, the process is actually easier than you might initially think. To set the arguments used for this section, simply type **:set args Arg1 Arg2** and press Enter. You can remove command line arguments using the :unset command.

To access the command-line arguments, you type **import System.Environment as Se** and press Enter. System.Environment contains the same sorts of functions found in other languages, as described at http://hackage.haskell.org/package/base-4.11.1.0/docs/System-Environment.html. For this example, you use only getArgs. To see the arguments you just provided, you can type **getArgs** and press Enter. You see a list containing the two arguments.

TIP

Obtaining a list of arguments means that you can process them using any of the list methods found earlier in this book and online. However, Chapter 11 also shows how to use monad sequencing, which works fine in this case by using the following code:

```
getArgs >>= mapM_ putStrLn
```

The output you see is each of the arguments displayed separately, one on each line, as shown in Figure 12-1. Of course, you could just as easily use a custom function to process the arguments in place of putStrLn. The tutorial at https://wiki.haskell.org/Tutorials/Programming_Haskell/Argument_handling gives you a better idea of how to use this approach with a custom parser.

FIGURE 12-1:
Haskell provides native techniques for accessing command line arguments.

TECHNICAL STUFF

When using the downloadable source for these examples, you still need to provide a command-line argument. However, using the :set command won't help. Instead, you need to type **:main Arg1 Arg2** and press Enter to get the same result after loading the code. Likewise, when working through the CmdArgs example found in the "Getting a simple command line in Haskell," later in this chapter, you type **:main --username=Sam** (with two dashes) and press Enter to obtain the correct result.

Making sense of the variety of packages

Haskell lacks any sort of command-line processing other than the native capability described in the previous section. However, you can find a wide variety of packages that provide various kinds of command-line argument processing on the Command Line Option Parsers page at https://wiki.haskell.org/Command_line_option_parsers. As mentioned on the page, the two current favorites are CmdArgs and optparse-applicative. This book uses the CmdArgs option (http://hackage.haskell.org/package/cmdargs) because it provides the simplest command-line parsing, but working with the other packages is similar.

If you need extensive command-line processing functionality, optparse-applicative (http://hackage.haskell.org/package/optparse-applicative) may be a better option, but it does come with some substantial coding requirements.

The multi-mode column on the Command Line Option Parsers page simply tells you how the Cabal (the Haskell installer) package is put together. Using a multi-mode package is more convenient because you need only one library to do everything, but many people go with the Linux principle of having a single task assigned to each library so that the library can do one thing and do it well.

Of more importance are the extensions and remarks columns for each package that appear on the Command Line Option Parsers page. The extensions describe the kinds of support that the package provides. For example, optparse-applicative supports the General Algebraic Datatypes (GADT) provided by Haskell (as described at https://en.wikibooks.org/wiki/Haskell/GADT). CmdArgs provides an extensive list of extensions, only three of which appear in the table. The remarks tell you about potential package issues, such as the lack of specific error messages for the Applicative Functor in optparse-applicative. The unsafePerformIO reference for CmdArgs refers to the method used to process code with side effects as describedathttp://hackage.haskell.org/package/base-4.11.1.0/docs/System-IO-Unsafe.html.

Obtaining CmdArgs

Before you can use CmdArgs, you must install it. The easiest way to do this is to open a command or terminal window on your system, type **cabal update**, and press Enter. This command ensures that you have the latest package list. After the update, type **cabal install cmdargs** and press Enter. Cabal will display a list of installation steps. Figure 12-2 shows the output you see in most cases.

When working with CmdArgs, you also see references to `DeriveDataTypeable`, which you can add to the top of your executable code by typing {-# **LANGUAGE DeriveDataTypeable #-}**. However, when working in the WinGHCi interpreter, you need to do something a bit different, as described in the following steps:

1. **Choose File ⇨ Options.**

   You see the dialog box shown in Figure 12-3.

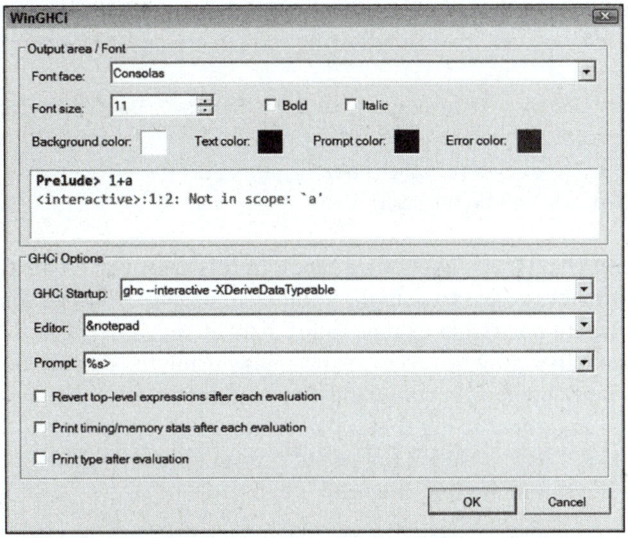

FIGURE 12-3: Add Derive-DataTypeable support to your interpreter.

2. **Add** -XDeriveDataTypeable **to the GHCi Startup field.**

   This option adds the required support to your interpreter. Don't remove any other command-line switches that you find in the field.

3. **Restart the interpreter.**

   You're ready to use CmdArgs.

OVERCOMING THE CABAL UPDATE ERROR

You may encounter an update error when attempting to update Cabal using `cabal update`. In this case, you can try `cabal --http-transport=plain-http update` instead. The problem is that Cabal is unable to resolve error messages from some sites.

Getting a simple command line in Haskell

Using a third-party library rather than cooking your own command-line parser has some specific advantages, depending on the library you use. This section discusses a minimum sort of command line, but you can use the information to make something more extensive. Before you can do anything, you need to add CmdArgs support to your application by typing **import System.Console.CmdArgs as Ca** and pressing Enter. You also need to set an argument for testing by typing **:set args --username=Sam** and pressing Enter. Make sure that you have no spaces in the argument and that you use two dashes, not one. Now that you have the support included, you can use the following code to create a test scenario.

```
data Greet = Greet {username :: String} deriving (Show,
    Data, Typeable)
sayHello = Greet {username = def}
print =<< cmdArgs sayHello
```

Chapter 10 tells you about data types. In this case, you create the `Greet` data type that provides access to a single argument, `username`, of type `String`. The next step is to create a variable of type `Greet` named `sayHello`. This is actually a kind of template that provides access to `username` using the default (`def`) arguments. The final line obtains the command-line argument using `cmdArgs` and formats any `--name` argument using the `sayHello` template. In this case, the output is `Greet {name = "Sam"}`. Notice the use of monad sequencing (`=<<`) to obtain the value from the command line and send it to `print`.

You'll want to do more than simply print the command line, which means accessing the values in some way. Chapter 10 showed how to perform a conversion of a custom type to a standard type using the `cvtToTuple` function. This example performs a similar conversion using the following code:

```
cvtToName (Greet {username=a}) = a
theName <- cmdArgs sayHello
putStrLn ("Hello " ++ (cvtToName theName))
```

The `cvtToName` function accepts a `Greet` object with a `name` and returns the string value that it contains. When you compare this function with `cvtToTuple` in the "Parameterizing Types" section of Chapter 10, you see that they're much alike in pattern.

WARNING

The next line may be a bit of a puzzle at first until you try typing **:t (cmdArgs sayHello)** and pressing Enter. The result is (`cmdArgs sayHello`) `:: IO Greet`, which isn't a `Greet` type, but rather an `IO Greet` type. Be sure to remember that Haskell relies on monads for I/O, as described in Chapter 11; a common mistake is to forget that you must deal with the results of using the IO operator to obtain access to the command-line arguments. When you obtain the type of theName, you find that it's of type Greet, which is precisely what you need as input to cvtToName.

The final line of code shows the complete conversion and output to screen using putStrLn. You could use this technique to obtain the value for any purpose. The CmdArgs main page shows you considerably more about displaying help information in various ways using the library. For example, it comes with --help and --version command-line switches by default.

Accessing the Command Line in Python

The Python command line is more traditional in most respects. As previously stated, it does make use of the functionality supplied by the operating system, as does every other language around, to obtain the command line. However, Python provides two forms of built-in support, with the Argparse library being favored for complex command-line management.

REMEMBER

The following sections give you a brief overview of the Python approach. Because Jupyter Notebook doesn't provide a convenient method of adding arguments to the command line, you need to rely on the Python interpreter instead. To access the Python interpreter, open the Anaconda Prompt (choose Start ⇨ All Programs ⇨ Anaconda3 on Windows systems and find it in the Anaconda3 folder).

Using the Python environment directly

The native Python command-line argument functionality follows that used by many other languages. For example, the information appears within argv, which is the same variable name used by languages such as C++. The following code shows typical access of argv from an application.

```python
import sys

print(sys.argv)
print(len(sys.argv))
if (len(sys.argv) > 0):
    print(sys.argv[0])
```

To test this script, type **python Native.py name=Sam** at the Anaconda prompt and press Enter. The output should show two arguments: Native.py and name=Sam. The command-line arguments always include the name of the application as the first argument. You can find additional information about using the native functionality at `http://www.pythonforbeginners.com/system/python-sys-argv` and `https://www.tutorialspoint.com/python/python_command_line_arguments.htm`.

Interacting with Argparse

Argparse provides some native functionality along the same lines as CmdArgs for Haskell. However, in this case, all you get is the -h command-line switch for help. Of course, just getting a help switch is nice, but hardly worthwhile for your application. The following code shows how to use Argparse to obtain a name and then display a hello message as output. Before you can do anything, you need to `import argparse` into the Python environment.

```
import argparse

parser = argparse.ArgumentParser()
parser.add_argument("name")
args = parser.parse_args()

nameStr = args.name.split("=")

print("Hello " + nameStr[1])
```

The first three lines of actual code create a parser, add an argument to it for `name`, and then obtain the list of arguments. When a user asks for help, `name` will appear as a positional argument.

You can access each argument by name, as shown in the next line of code. The argument will actually appear as `name=Sam` if you supply Sam as the name at the command line. The combination of the two elements isn't useful, though, so the example splits the string at the = sign. Finally, the example outputs the message with the supplied name. You can test this example by typing **python Argparse.py name=Sam** and pressing Enter at the command line.

This example was just enough to get you started and to demonstrate that Python also provides a great library with added command-line functionality. You can find out more about Argparse at `https://docs.python.org/3/howto/argparse.html`

Chapter **13**

Dealing with Files

C hapter 11 gives you a very brief look at localized file management in the "Working with devices" section of the chapter. Now it's time to look at local files in more detail because you often use local files as part of applications — everything from storing application settings to analyzing a moderately large dataset. In fact, as you may already know, local files were the first kind of data storage that computers used; networks and the cloud came much later. Even on the smallest tablet today, you can still find local files stored in a hard-drive–like environment (although hard drives have come a very long way from those disk packs of old).

After you get past some of the general mechanics of how files are stored, you actually need to start working with them. Developers face a number of issues when working with files. For example, one of the more common problems is that a user can't access a file because of a lack of rights. Security is a two-edged sword that protects data by restricting access to it and keeping the right people from accessing it for the right reasons. This chapter helps you understand various file access issues and demonstrates how to overcome them.

The chapter also discusses Create, Read, Update, and Delete (CRUD), the four actions you can perform on any file for which you have the correct rights. CRUD normally appears in reference to database management, but it applies just as much to any file you might work with.

Understanding How Local Files are Stored

If you have worked with computers for a while, you know that the operating system handles all the details of working with files. An application requests these services of the operating system. Using this approach is important for security reasons, and it ensures that all applications can work together on the same system. If each application was allowed to perform tasks in a unique manner, the resulting chaos would make it impossible for any application to work.

The reason that operating system and other application considerations are important for the functional programming paradigm is that unlike other tasks you might perform, file access depends on a nonfunctional, procedural third party. In most cases, you must perform a set of prescribed steps in a specific order to get any work done. As with anything, you can find exceptions, such as the functional operating systems described at `http://wiki.c2.com/?PurelyFunctional OperatingSystem` and `https://en.wikipedia.org/wiki/House_(operating_system)`. However, you have to ask yourself whether you've ever even heard of these operating systems. You're more likely to need to work with OS X, Linux, or Windows on the desktop and something like Android or iOS on mobile devices.

REMEMBER

Most operating systems use a hierarchical approach to storing files. Each operating system does have differences, such as those discussed between Linux and Windows at `https://www.howtogeek.com/137096/6-ways-the-linux-file-system-is-different-from-the-windows-file-system/`. However, the fact that Linux doesn't use locks on files but Windows does really won't affect your application in most cases. The recursive nature of the functional programming paradigm does work well in locating files and ensuring that files get stored in the right location. Ultimately, the hierarchy used to store files means that you need a path to locate the file on the drive (regardless of whether the operating system specifically mentions the drive).

Files also have specific characteristics associated with them that vary by operating system. However, most operating systems include a creation and last modification date, file size, file type (possibly through the use of a particular file extension), and security access rights with the filename. If you plan to use your application on multiple platforms, which is becoming more common, you must create a plan for interacting with file properties in a consistent manner across platforms if possible.

All the considerations described in this section come into play when performing file access, even with a functional language. However, as you see later, functional languages often rely on the use of monads to perform most file access tasks in a consistent manner across operating systems, as described for any I/O in Chapter 11. By abstracting the process of interacting with files, the functional programming paradigm actually makes things simpler.

Ensuring Access to Files

A number of common problems arise in accessing files on a system — problems that the functional programming paradigm can't hide. The most common problem is a lack of rights to access the file. Security issues plague not only the local drive, but every other sort of drive as well, including cloud-based storage. One of the best practices for a developer to follow is to test everything using precisely the same rights that the user will have. Unfortunately, even then you may not find every security issue, but you'll find the vast majority of them.

Some access issues are also the result of bad information — fallacies that developers have simply believed without testing. One of these issues is the supposed difference in using the backslash on Windows and the forward slash on Linux and OS X. The truth is that you can use the forward slash on all operating systems, as described at http://blog.johnmuellerbooks.com/2014/03/10/backslash-versus-forward-slash/. All the example code in this chapter uses the forward slash when dealing with paths as a point of demonstration.

Often a developer also runs afoul of file property issues. Some of these issues are external to the file, such as mistaking one file type for another. Other issues are internal to the file, such as trying to read a UTF-7 file using code designed for UTF-8 or UTF16, which are currently more common. Even though you can access a file when facing a property issue, the access doesn't help because you can't do anything with the file after you access it. As far as your application is concerned, you still lack access to the file (and in a practical sense, you do, even if you have successfully opened it).

Specific language tools also present problems. For example, the message thread at https://github.com/haskell/cabal/issues/447 discusses issues that occur as part of the installation process using Cabal (the utility that ships with Haskell). Imagine installing a new application that you built and then finding that only administrators can use it. Unfortunately, this problem might not show up unless you test your application installation on the right version of Windows. Haskell isn't alone in this problem; every language comes with special issues that may affect your ability to access files, so constant testing and handling of error reports is an essential part of working with files.

Interacting with Files

Understanding how the files are stored and knowing the requirements for access are the first two steps in interacting with them. If you have worked with other programming languages, you have likely worked with files in a procedural manner:

obtaining a file handle, using it to open the file, and then closing the file handle when finished. The functional programming paradigm must also follow these rules, as demonstrated in Chapter 11, but working in the functional world brings different nuances, as discussed in the sections that follow.

Creating new files

Operating systems generally provide a number of ways of opening files. In the default method, you normally open the file and overwrite the existing content with anything new that you write. When the file doesn't exist, the operating system automatically creates it for you. The following code shows an example of opening a file for writing and automatically creating that file when it doesn't exist:

```
import System.IO as IO

main = do
    handle <- openFile "MyData.txt" WriteMode
    hPutStrLn handle "This is some test data."
    hClose handle
```

The defining factor here is the WriteMode argument. When you use the WriteMode argument, you tell the operating system to create a new file when one doesn't exist or to overwrite any existing content. The Python equivalent to this code is

```
handle = open("MyData2.txt", "w")
print(handle.write("This is some test data.\n"))
handle.close()
```

Notice that when using Python, you use the "w" argument to access the write mode. In addition, Python has no method of writing a line with a carriage return; you add it manually by using the \n escape. Adding the print function lets you see how many characters Python writes to the file.

REMEMBER

As an alternative to using the WriteMode argument, you can use the ReadWriteMode argument when you want to both read from and write to the file. Writing to the file works as before: You either create a new file or overwrite the content of an existing file. To read from the file, of course, the file must contain something to read. The "Reading data" section of the chapter discusses this issue in more detail.

FILE LOCKING OVERVIEW

When working with data files, it's generally important to perform a complete or partial lock of the file while the data changes or you risk overwriting the data. Databases normally use record level locks so that several people can work with the file at the same time.

Depending on your operating system, however, you may find that the operating system doesn't lock files — or at least not with an actual lock (see https://www.howtogeek.com/141393/why-cant-i-alter-in-use-files-on-windows-like-i-can-on-linux-and-os-x/ and https://stackoverflow.com/questions/196897/locking-executing-files-windows-does-linux-doesnt-why for details). In addition, some applications actually follow a policy of not locking the file and prefer using a "last edit wins" approach to dealing with data changes.

Sometimes rules like file locking can actually cause problems. Articles like the one at https://success.outsystems.com/Documentation/10/Developing_an_Application/Use_Data/Offline/Offline_Data_Sync_Patterns/Read%2F%2FWrite_Data_Last_Write_Wins describe why this approach is actually beneficial when working with mobile applications. The programming language you use may also change how file locks work, with many languages automatically incorporating file locking unless you specify otherwise. The bottom line is to know whether file locking occurs with your operating system and language combination, determine when file locking is beneficial, and set a policy that specifically defines file locking for your application.

Opening existing files

When you have an existing file, you can read, append, update, and delete the data it contains. Even though you will create new files when writing an application, most applications spend more time opening existing files in order to manage content in some way. For the application to perform data-management tasks, the file must exist. Even if you think that the file exists, you must verify its presence because the user or another application may have deleted it, or the user may not have followed protocol and created it, or sunspot could have damaged the file directory entry on disk, or . . .. The list can become quite long as to why the file you thought was there really isn't. The process of data management can become complex because you often perform searches for specific content as well. However, the initial task focuses on simply opening the file.

The "Reading data" section of the chapter discusses the task of opening a file to read it, especially when you need to search for specific data. Likewise, writing, updating, and deleting data appears in the "Updating data" section of the chapter.

However, the task of *appending* — adding content to the end of the file — is somewhat different. The following code shows how to append data to a file that already exists:

```
import System.IO as IO

main = do
    handle <- openFile "MyData.txt" AppendMode
    hPutStrLn handle "This is some test data too."
    hClose handle
```

Except for the AppendMode argument, this code looks much like the code in the previous section. However, no matter how often you run the code in the previous section, the resulting file always contains just one line of text. When you run this example, you see multiple lines, as shown in Figure 13-1.

FIGURE 13-1:
Appending means adding content to the end of a file.

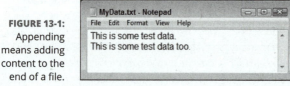

Python provides the same functionality. The following code shows the Python version, which relies on the "a" (append) mode:

```
handle = open("MyData2.txt", "a")
print(handle.write("This is some test data too.\n"))
handle.close()
```

REMEMBER

Some languages treat appending differently from standard writing. If the file doesn't exist, the language will raise an exception to tell you that you can't append to a file that doesn't exist. To append to a file, you must create it first. Both Haskell and Python take a better route — appending also covers creating a new file when one doesn't exist.

Manipulating File Content

When thinking through the process of dealing with I/O on the local system in the form of files, you have to separate the main components and deal with them individually:

>> **Physicality:** The location of the file on the storage system. The operating system can hide this location in some respects, and even create mappings so that a single storage unit actually points to multiple physical drives that aren't necessarily located on the local machine. The fact remains, however, that the file must appear somewhere. Even if the user accesses this file by clicking a convenient icon, the developer must still have some idea of where the file resides, or access is impossible.

>> **Container:** Data resides in a container of some sort. The container used in this chapter is a file, but it could just as easily be a database or a collection of files within a particular folder. As with physicality, users don't often see the container used to hold the data except as an abstraction (and sometimes not even that, as in the case of an application that opens a database automatically). Again, the developer must know the properties and characteristics of the container to write a successful application.

>> **Data:** The data itself is an entity and the one that everyone, including users, is intimately aware of when working with an application. Previous sections of the chapter discuss the other entities in this list. The following sections discuss this final entity. It begins with the Create, Read, Update, and Delete (CRUD) operations associated with data and views two of those entities in closer detail.

Considering CRUD

People create acronyms to make remember something easier. Sometimes those acronyms are unfortunate, as in calling operations on data CRUD. However, the people who work with databases wanted something easy to remember, so data-related tasks became CRUD. Another school of thought called the list of tasks Browse, Read, Edit, Add, and Delete (BREAD), but that particular acronym didn't seem to stick, even though your daily BREAD might rely on your ability to employ CRUD. This chapter uses CRUD because that seems to be the most popular acronym. You can view CRUD as comprising the following tasks:

>> **Create:** Adding new data to storage. Anytime you create new storage, such as a file, you generally create new data as well. Empty storage isn't useful. The examples in the "Interacting with Files" section, earlier in this chapter, demonstrate creating data in both a new and an existing file. In both cases, the functional programming paradigm uses the IO monad operation on the combination of a handle and the associated data to place data in the file. This takes place after creating the file using another monad consisting of the IO operating on a combination of the filename and opening mode.

» **Read:** Reading data within a storage container means to do something with the content that doesn't change it in any way. You can see at least two kinds of read tasks in most applications:

a. Employ an IO monad operation on the combination of a handle and data location to retrieve specific data. In this case, the data output is the target of the task. When you don't supply a specific location, the operation assumes either the start of the storage or the current storage location pointer value. (The *location pointer* is an internally maintained value that indicates the end of the last read location within the storage.)

b. Employ an IO monad operating on the combination of a handle and search criteria. In this case, the goal is to search for specific data and retrieve a data location based on that search. Some developers view this task as a browse, rather than as a read.

» **Update:** When data within the storage container still has value but contains mistakes, it requires an update, which the application performs using the following steps. In this case, you're really looking at a series of IO monads:

1. Locate the existing data using the combination of a handle and the search expression.

2. Copy the existing data using the combination of a handle and the data location.

3. Write the new data using a combination of a handle and the data.

» **Delete:** When the data within storage no longer has value, the application deletes the entry. In this case, you rely on the following IO monads to perform the task:

1. Locate the data to remove using a combination of a handle and a search expression.

2. Delete the data using a combination of a handle and a data location.

Reading data

The concept of reading data isn't merely about obtaining information from a storage container, such as a file. When a person reads a book, a lot more goes on than simple information acquisition, in many cases. Often, the person must search for the appropriate information (unless the intent is to read the entire book) and then track progress during each reading session (unless there is just one session). A computer must do the same. The following example shows how the computer tracks its current position within the file during the read:

```
import System.IO as IO

main = do
    handle <- openFile "MyData.txt" ReadMode
    myData <- hGetLine handle
    position <- hGetPosn handle
    hClose handle
    putStrLn myData
    putStrLn (show position)
```

Here, the application performs a read using hGetLine, which obtains an entire line of text (ending with a carriage return). However, the test file contains more than one line of text if you worked through the examples in the previous sections. This means that the file pointer isn't at the end of the file.

The call to hGetPosn obtains the actual position of the file pointer. The example outputs both the first line of text and the file position, which is reported as {handle: MyData.txt} at position 25 if you used the file from the previous examples. A second call to hGetLine will actually retrieve the next line of text from the file, at which point the file pointer will be at the end of the file.

REMEMBER

The example shows hGetLine, but Haskell and Python both provide an extensive array of calls to obtain data from a file. For example, you can get a single character by calling hGetChar. You can also peek at the next character in line without moving the file pointer by calling hLookAhead.

Updating data

Of the tasks you can perform with a data container, such as a file, updating is often the hardest because it involves finding the data first and then writing new data to the same location without overwriting any data that isn't part of the update. The combination of the language you use and the operating system do reduce the work you perform immensely, but the process is still error prone. The following code demonstrates one of a number of ways to change the contents of a file. (Note that the two lines beginning with let writeData must appear on a single line in your code file.)

```
import System.IO as IO
import Data.Text as DT

displayData (filePath) = do
    handle <- openFile filePath ReadMode
    myData <- hGetContents handle
```

```
    putStrLn myData
    hClose handle

main = do
    displayData "MyData3.txt"

    contents <- readFile "MyData3.txt"
    let writeData = unpack(replace
        (pack "Edit") (pack "Update") (pack contents))
    writeFile "MyData4.txt" writeData

    displayData "MyData4.txt"
```

This example shows two methods for opening a file for reading. The first (as defined by the displayData function) relies on a modified form of the code shown in the "Reading data" section, earlier in this chapter. In this case, the example gets the entire contents of the file in a single read using hGetContents. The second version (starting with the second line of the main function) uses readFile, which also obtains the entire content of the file in a single read. This second form is easier to use but provides less flexibility.

The code uses the functions found in Data.Text to manipulate the file content. These functions rely on the Text data type, not the String data type. To convert a String to Text, you must call the pack function, as shown in the code. The reverse operation relies on the unpack function. The replace function provides just one method of modifying the content of a string. You can also rely on mapping to perform certain kinds of replacement, such as this single-character replacement:

```
let transform = pack contents
DT.map (\c -> if c == '.' then '!' else c) transform
```

This method relies on a lambda function and provides considerable flexibility for a single-character replacement. The output replaces the periods in the text with exclamation marks by mapping the lambda function to the packed String (which is a Text object) found in transform. Notice how the lambda function examines characters separately, as opposed to the word-level search used in the example.

WARNING

Observe how the example uses one file for input and an entirely different file for output. Haskell relies on lazy reads and writes. If you were to attempt to use readFile on a file and then writeFile on the same file a few lines down, the resulting application would display a "resource busy" type of error message.

Completing File-related Tasks

After you finish performing data-related tasks, you need to do something with the data storage container. In most cases, that means closing the handle associated with the container. When working with files, some functions, such as readFile and writeFile, perform the task automatically. Otherwise, you close the file manually using hClose.

WARNING

Haskell, like most languages, comes with a few odd calls. For example, when you call hGetContents, the handle you use is semi-closed. A semi-closed handle is almost but not quite closed, which is odd when you think about it. You can't perform any additional reads, nor can you obtain the position of the file pointers. However, calling hClose to fully close the handle is still possible. The odd nature of this particular call can cause problems in your application because the error message will tell you that the handle is semi-closed, but it won't tell you what that means or define the actual source of the semi-closure.

Another potential need may arise. If you use temporary files in your application, you need to remove them. The removeFile function performs this task by deleting the file from the path you supply. However, when working with Haskell, you find the call in System.Directory, not System.IO.

Chapter **14**

Working with Binary Data

The term binary data is an oxymoron because as far as the computer is concerned, only binary data exists. *Binary data* is the data that people associate with a nonhuman-readable form; the data is a series of seemingly unrelated 0s and 1s that somehow form patterns the computer sees as data, despite the human inability to do so in many cases — at least, not without analysis. Consequently, when this chapter contrasts textual data to binary data, it does so from the human perspective, which means that data must be readable and understandable by humans to be meaningful. Of course, with computer assistance, binary data is also quite meaningful, but in a different way from text. This chapter begins by helping you understand the need and uses for binary data.

The days of worrying about data usage at the bit level are long gone, but binary data, in which individual bits do matter, still appears as part of data analysis. The search for patterns in data isn't limited to human-readable form, nor is the output from an analysis always in human-readable form, even when the input is. Consequently, you need to understand the role of the binary form in data analysis. As part of understanding why functional programming is so important, this chapter considers the use of binary data in data analysis.

Binary data also appears in many human-pleasing forms. For example, raster graphic files rely exclusively on binary data for the data-storage part of the file. The conversion of a human-readable file to a compressed form also appears as binary data until you decompress it. This chapter explores a few of these forms of binary data. The chapter doesn't explore binary file forms in any depth, but you do get an overview of them.

Comparing Binary to Textual Data

Chapter 13 discusses textual data. All the information in that chapter is in a human-readable form. Likewise, most data you encounter directly today is in some human-readable form, much of it textual. However, under the surface lies the binary data that the computer understands. The true difference between binary and textual data is interpretation — that is, how humans see the data (or don't see it). The letter A is simply the number 65 in disguise when viewed as ASCII.

Oddly enough, the ASCII numeric representation of the letter A isn't the end of the line. Somewhere, a raster representation of the letter A exists that determines what you see as the letter A in print or onscreen. (The article at https://www.ibm.com/support/knowledgecenter/en/SSLTBW_2.3.0/com.ibm.zos.v2r3.e0zx100/e0z2o00_char_rep.htm discusses raster representations in more detail.) The fact is that the letter A doesn't actually exist in your computer; you simply see a representation of it that is quite different to the computer.

REMEMBER

When it comes to numeric data, the whole issue of textual versus binary data becomes more complex. The number could appear as text — meaning a sequence of characters expressing a numeric value. The ASCII values 48 through 57 provide the required textual values. Add a decimal point and you have a human-readable, textual number.

However, a numeric value can also appear as a number in various forms that the computer will directly understand (integers) or require to be translated (as in the IEEE 754 floating-point values). Even though integers and floating-point values both appear as 0s and 1s, the human interpretation often differs from the computer interpretation. For example, in a single-precision floating-point value, the computer sees 32-bits of data — just a series of 0s and 1s that mean nothing to the computer. Yet the interpretation requires splitting those bits into one sign bit, 8 exponent bits, and 23 significand bits (see https://www.geeksforgeeks.org/floating-point-representation-basics/ for details).

REMEMBER

Underlying all these representations of data that humans create is a binary stream that the computer controls and understands. All the computer sees is 0s and 1s. The computer merely manipulates the stream and, as with any other machine, has no understanding whatsoever of what those 0s and 1s mean. When you work with binary data, what you really do is work with the computer presentation of the human-readable form that you want to express, no matter what form that data may take. All data is binary to the computer. To make the data useful, however, an application must take the binary presentation and translate it in some way to create a form that humans can understand and use.

Also, a particular language makes specific presentations available and controls the manner in which you create and manipulate the presentations. However, no language actually controls the underlying data, which is always in 0s and 1s. A language interacts with the underlying data through libraries, the operating system, and the machine hardware itself. Consequently, all languages share the same underlying data type, which is binary.

Using Binary Data in Data Analysis

Binary data figures strongly in data analysis, where it often indicates a Boolean value — that is, True or False. Some languages use an entire *byte* (8 bits in most cases) or even a *word* (16, 32, or 64bits, in most cases) to hold Boolean values because memory is cheap and manipulating individual bits can be time consuming. However, other languages use each bit in a byte or word to indicate truth-values in a form called *flags*. A few languages provide both options.

The Boolean value often indicates the outcome of a data analysis such as a Bernoulli trial (see `http://www.mathwords.com/b/bernoulli_trials.htm` for details). In pure functional programming languages, a Bernoulli trial is often expressed as a binomial distribution (see `https://hackage.haskell.org/package/statistics-0.14.0.2/docs/Statistics-Distribution-Binomial.html`) and the language often provides specific functionality to perform the calculations. When working with impure languages, you can either simulate the effect or rely on a third-party library for support, such as NumPy (see `https://docs.scipy.org/doc/numpy/reference/generated/numpy.random.binomial.html` for details). The example at `http://www.chadfulton.com/topics/bernoulli_trials_classical.html` describes the specifics of performing a Bernoulli trial in Python.

TECHNICAL STUFF

When considering binary data, you need to think about how the calculation you perform can skew any results obtained. For example, many people use the coin toss as an example for explaining the Bernoulli trial. However, it works only when you ignore the possibility of a coin landing on its edge, landing on neither heads or tails. Even though the probability of such a result is incredibly small, a true analysis would consider it a potential output. However, to calculate the result, you must now eschew the use of binary analysis, which would greatly increase calculation times. The point is that data analysis is often an imperfect science, and the person performing the required calculations needs to consider the ramifications of any shortcuts used in the interest of speed.

Of course, Boolean values (binary data, really) is used for Boolean algebra (see http://mathworld.wolfram.com/BooleanAlgebra.html for details) where the truth value of a particular set of expressions comes as a result of the logical operators applied to the target monads. In many cases, the outcome of such binary analysis sees visual representation as a Hasse diagram (see http://mathworld.wolfram.com/HasseDiagram.html) for details.

Every computer language today has built-in primitives for performing Boolean algebra. However, pure functional languages also have libraries for performing more advanced tasks, such as the Data.Algebra.Boolean Haskell library discussed at http://hackage.haskell.org/package/cond-0.4.1.1/docs/Data-Algebra-Boolean.html. As with other kinds of analysis of this sort, impure languages often rely on third-party libraries, such as the SymPy library for Python discussed at http://docs.sympy.org/latest/modules/logic.html.

This section could easily spend more time on data analysis, but one final consideration is regression analysis of binary variables. Regression analysis takes in a number of analysis types, some of which appear at https://www.analyticsvidhya.com/blog/2015/08/comprehensive-guide-regression/. The most common for binary data are logistic regression (see http://www.statisticssolutions.com/what-is-logistic-regression/) and probit regression (see https://stats.idre.ucla.edu/stata/dae/probit-regression/). Even in this case, pure functional languages tend to provide built-in support, such as the Haskell library found at http://hackage.haskell.org/package/regress-0.1.1/docs/Numeric-Regression-Logistic.html for logistic regression. Of course, third-party counterparts exist for impure languages, such as Python (see http://scikit-learn.org/stable/modules/generated/sklearn.linear_model.LogisticRegression.html).

Understanding the Binary Data Format

As mentioned in earlier sections, the computer manages binary data without understanding it in any way. Moving bits around is a mechanical task. In fact, even the concept of bits is foreign because the hardware sees only differences in voltage between a 0 and a 1. However, to be useful, the binary data must have a *format*; it must be organized in some manner that creates a pattern. Even text data of the simplest sort has formatting that defines a pattern. One of the best ways to understand how this all works is to actually examine some files using a hexadecimal editor such as XVI32 (http://www.chmaas.handshake.de/delphi/freeware/xvi32/xvi32.htm). Figure 14-1 shows an example of this tool in action using the extremely simple MyData.txt file that you create in Chapter 13.

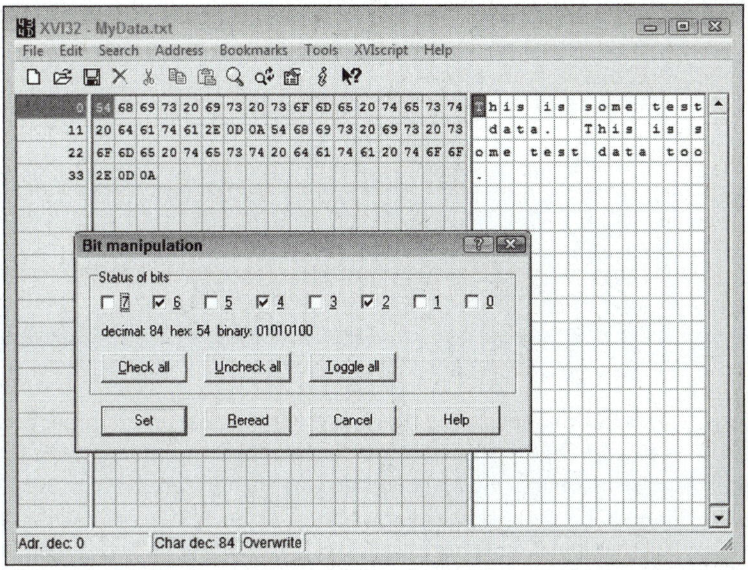

FIGURE 14-1:
Use a product such as XVI32 to understand binary better.

In this case, you see the hexadecimal numbers in the middle pane of the main window and the associated letters in the right pane. The Bit Manipulation dialog box shows the individual bits used to create the hexadecimal value. What the computer sees is those bits and nothing more. However, in looking at this file, you can see the pattern—one character following the next to create words and then sentences. Each sentence ends with a 0D (carriage return) and a 0A (line feed). If you decided that it was in your best interest to do so, you could easily create this file using binary methods, but Chapter 13 shows the easier method of using characters.

Every file on your system has a format of some sort or it wouldn't contain useful information. Even executable files have a format. If you're working with Windows, many of your executables will rely on the MZ file format described at https://www.fileformat.info/format/exe/corion-mz.htm. Figure 14-2 shows the XVI32.exe executable file (just the bare beginning of it). Notice that the first two letters in the file are MZ, which identify it as an executable that will run under Windows. When a native executable lacks this signature, Windows won't run it unless it's part of some other executable format. If you follow the information found on the FileFormat.Info site, you can actually decode the content of this executable to learn more about it. The executable even contains human readable text that you can use to discover some additional information about the application.

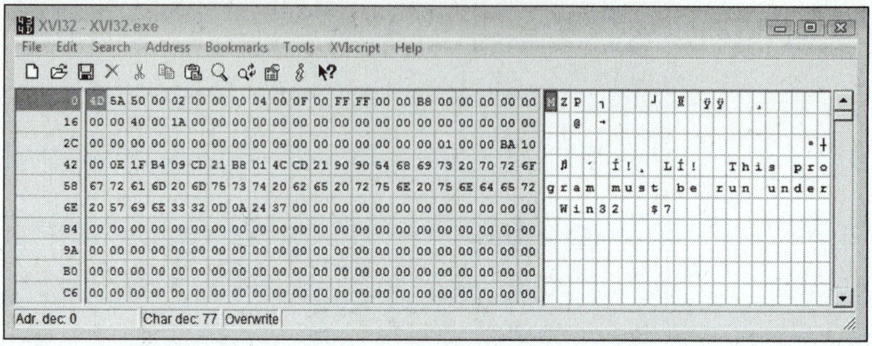

FIGURE 14-2:
Even executables
have a format.

REMEMBER

This information is important to the functional programmer because the languages (at least the pure ones) provide the means to interact with bits should the need arise in a mathematical manner. One such library is Data.Bits (http://hackage. haskell.org/package/base-4.11.1.0/docs/Data-Bits.html) for Haskell. The bit manipulation features in Haskell are somewhat better than those found natively in Python (https://wiki.python.org/moin/BitManipulation), but both languages also support third-party libraries to make the process easier. Given a need, you can create your own binary formats to store specific kinds of information, especially the result of various kinds of analysis that can rely on bit-level truth-values.

Of course, you need to remember the common binary formats used to store data. For example, a Joint Photographic Experts Group (JPEG) file uses a binary format (see https://www.fileformat.info/format/jpeg/internal.htm), which has a signature of JFIF (JPEG File Information Format), as shown in Figure 14-3. The use of this signature is similar to the use of the MZ for executable files. A study of the bits used for graphic files can consume a lot of time because so many ways exist to store the information (see https://modassicmarketing.com/understanding-image-file-types). In fact, so many storage methodologies are available for just graphic files that people have divided the formats into groups, such as lossy versus lossless and vector versus raster.

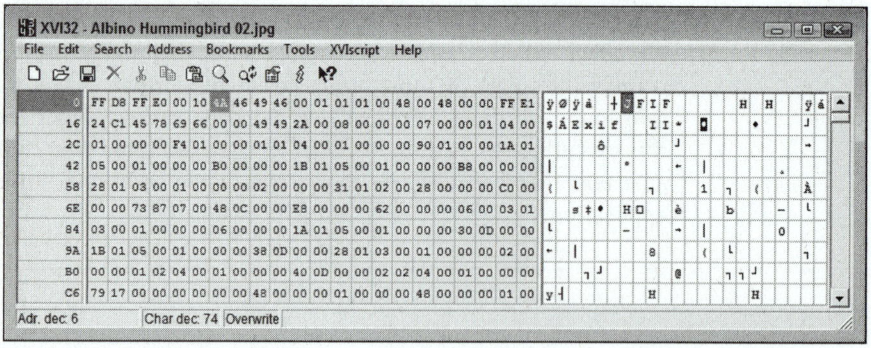

FIGURE 14-3:
Many binary
files include
signatures to
make them
easier to
identify.

Working with Binary Data

So far, this chapter has demonstrated that binary data exists as the only hardware-manipulated data within a computer and that binary data exists in every piece of information you use. You have also discovered that languages generally use abstractions to make the binary data easier to manipulate (such as by using text) and that functional languages have certain advantages when working directly with binary data. The question remains, however, as to why you would want to work directly with binary data when the abstractions exist. For example, you have no reason to create a JPEG file using bits when libraries exist to manipulate them graphically. A human understands the graphics, not the bits. In most cases, you don't manipulate binary data directly unless one of these conditions arises:

>> No binary format exists to store custom data containing binary components.

>> The storage capabilities of the target device have strict limits on size.

>> Transmitting data stored using less efficient methods is too time consuming.

>> Translating between common storage forms and the custom form needed to perform a task requires too much time.

>> A common storage format file contains an error that self-correction can't locate and fix.

>> You need to perform bit-level data transfers so that you can perform machine control, for example.

>> Curiosity mandates studying the file format in detail.

Interacting with Binary Data in Haskell

The examples presented in this section are extremely simple. You can find a considerable number of complex examples online; one appears at `http://hackage.haskell.org/package/bytestring-0.10.8.2/docs/Data-ByteString-Builder.html` and `https://wiki.haskell.org/Serialisation_and_compression_with_Data_Binary`. However, most of these examples don't answer the basic question of what you need to do as a minimum, which is what you find in the following sections. For these cases, you write several data types to a file, examine the file, and then read the data back using the simplest methods possible.

Writing binary data using Haskell

Remember that you have no limitations when working with data in binary mode. You can create any sort of output necessary, even concatenating unlike types together. The best way to create the desired output is to use `Builder` classes, which contain the tools necessary to build the output in a manner similar to working with blocks. The `Data.Binary.Builder` and `Data.ByteString.Builder` libraries both contain functions that you can use to create any needed output, as shown in the following code:

```
import Data.Binary.Builder as DB
import Data.ByteString.Builder as DBB
import System.IO as IO

main = do
   let x1 = putStringUtf8 "This is binary content."
   let y = putCharUtf8 '\r'
   let z = putCharUtf8 '\n'
   let x2 = putStringUtf8 "Second line..."

   handle <- openBinaryFile "HBinary.txt" WriteMode
   hPutBuilder handle x1
   hPutBuilder handle y
   hPutBuilder handle z
   hPutBuilder handle x2
   hClose handle
```

This example uses two functions, `putStringUtf8` and `putCharUtf8`. However, you also have access to functions for working with data types such as integers and floats. In addition, you have access to functions for working in decimal or hexadecimal as needed.

The process for working with the file is similar to working with a text file, but you use the `openBinaryFile` function instead to place Haskell in binary mode (where it won't interpret your data) versus text mode (where it does interpret things like escape characters). When outputting the values, you use the `hPutBuilder` function to chain them together. Putting output together like this (or using other, more complex methods) is called *serialization.* You serialize each of the outputs so that they appear in the file in the right order. As always, close the handle when you finish with it. Figure 14-4 shows the binary output of this application, which includes the carriage return and linefeed control characters.

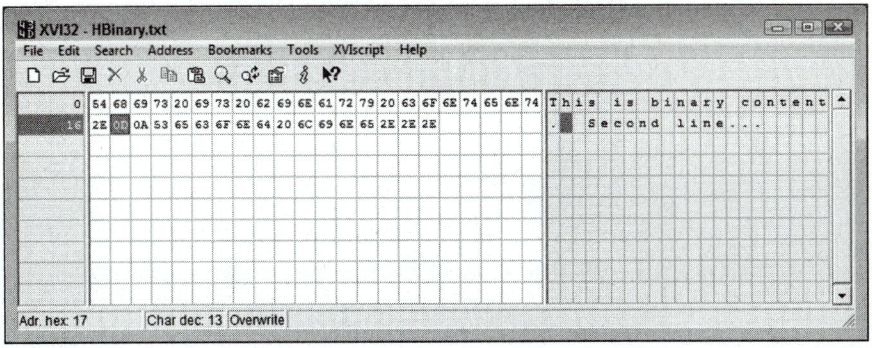

FIGURE 14-4: Even though this output contains text, it could contain any sort of data at all.

Reading binary data using Haskell

This example uses a simplified reading process because the example file does contain text. Even so, the `Data.ByteString.Char8` library contains functions for reading specific file lengths. This means that you can read the file a piece at a time to deal with different data types. The process of reading a file and extracting each of the constituent parts is called *deserialization.* The following code shows how to work with the output of this example in binary mode.

```
import Data.ByteString.Char8 as DB
import System.IO as IO

main = do
    handle <- openBinaryFile "HBinary.txt" ReadMode
    x <- DB.hGetContents handle
    DB.putStrLn x
    hClose handle
```

Notice that you must precede both `hGetContents` and `putStrLn` with `DB`, which tells Haskell to use the `Data.ByteString.Char8` functions. If you don't make this distinction, the application will fail because it won't be able to determine whether to use `DB` or `IO`. However, if you guess wrong and use `IO`, the application will still fail because you need the functions from `DB` to read the binary content. Figure 14-5 shows the output from this example.

FIGURE 14-5: The result of reading the binary file is simple text.

Interacting with Binary Data in Python

Python uses a more traditional approach to working with binary files, which can have a few advantages, such as being able to convert data with greater ease and having fewer file management needs. Remember that Haskell, as a pure language, relies on monads to perform tasks and expressions to describe what to do. However, when you review the resulting files, both languages produce precisely the same output, so the issue isn't one of how one language performs the task as contrasted to another, but rather which language provides the functionality you need in the form you need it. The following sections look at how Python works with binary data.

Writing binary data using Python

Python uses a lot of subtle changes to modify how it works with binary data. The following example produces precisely the same output as the Haskell example found in the "Writing binary data using Haskell" section, earlier in this chapter.

```
handle = open("PBinary.txt", "wb")
print(handle.write(b"This is binary content."))
print(handle.write(bytearray(b'\x0D\x0A')))
print(handle.write(b"Second line..."))
handle.close()
```

When you want to open a file for text-mode writing, in which case the output is interpreted by Python, you use "w". The binary version of writing relies on "wb", where the b provides binary support. Creating binary text is also quite easy; you simply prepend a b to the string you want to write. An advantage to writing in binary mode is that you can mix bytes in with the text by using a type such as bytearray, as shown in this example. The \x0D and \x0A outputs represent the carriage return and newline control characters. Of course, you always want to close the file handle on exit. The output of this example shows the number of bytes written in each case:

```
23
2
14
```

Reading binary data using Python

Reading binary data in Python requires conversion, just as it does in Haskell. Because this example uses pure text (even the control characters are considered text), you can use a simple decode to perform the task, as shown in the following code. Figure 14-6 shows the output of running the example.

```python
handle = open("PBinary.txt", "rb")
binary_data = handle.read()
print(binary_data)
data = binary_data.decode('utf8')
print(data)
```

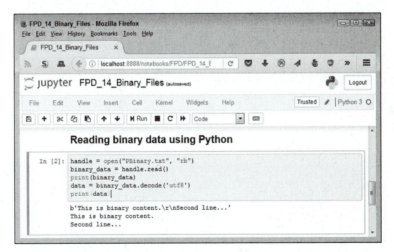

FIGURE 14-6:
The raw binary data requires decoding before displaying it.

Chapter **15**

Dealing with Common Datasets

The reason to have computers in the first place is to manage data. You can easily lose sight of the overriding goal of computers when faced with all the applications that don't seem to manage anything. However, even these applications manage data. For example, a graphics application, even if it simply displays pictures from last year's camping trip, is still managing data. When looking at a Facebook page, you see data in myriad forms transferred over an Internet connection. In fact, it would be hard to find a consumer application that doesn't manage data, and impossible to find a business application that doesn't manage data in some way. Consequently, data is king on the computer.

REMEMBER

The datasets in this chapter are composed of a specific kind of data. For you to be able to perform comparisons, conduct testing, and verify results of a group of applications, each application must have access to the same standard data. Of course, more than just managing data comes into play when you're considering a standard dataset. Other considerations involve convenience and repeatable results. This chapter helps you take these various considerations into account.

Because the sorts of management an application performs differs by the purpose of the application, the number of commonly available standard datasets is quite large. Consequently, finding the right dataset for your needs can be time consuming. Along with defining the need for standardized datasets, this chapter also looks at methods that you can use to locate the right standard dataset for your application.

After you have a dataset loaded, you need to perform various tasks with it. An application can perform a simple analysis, display data content, or perform Create, Read, Update, and Delete (CRUD) tasks as described in the "Considering CRUD" section of Chapter 13. The point is that functional applications, like any other application, require access to a standardized data source to look for better ways of accomplishing tasks.

Understanding the Need for Standard Datasets

A *standard dataset* is one that provides a specific number of records using a specific format. It normally appears in the public domain and is used by professionals around the world for various sorts of tests. Professionals categorize these datasets in various ways:

>> Kinds of fields (features or attributes)

>> Number of fields

>> Number of records (cases)

>> Complexity of data

>> Task categories (such as classification)

>> Missing values

>> Data orientation (such as biology)

>> Popularity

Depending on where you search, you can find all sorts of other information, such as who donated the data and when. In some cases, old data may not reflect current social trends, making any testing you perform suspect. Some languages actually build the datasets into their downloadable source so that you don't even have to do anything more than load them.

WARNING

Given the mandates of the General Data Protection Regulation (GDPR), you also need to exercise care in choosing any dataset that could potentially contain any individually identifiable information. Some people didn't prepare datasets correctly in the past, and these datasets don't quite meet the requirements. Fortunately, you have access to resources that can help you determine whether a dataset is acceptable, such as the one found on IBM at `https://www.ibm.com/security/data-security/gdpr`. None of the datasets used in this book are problematic.

Of course, knowing what a standard dataset is and why you would use it are two different questions. Many developers want to test using their own custom data, which is prudent, but using a standard dataset does provide specific benefits, as listed here:

>> Using common data for performance testing

>> Reducing the risk of hidden data errors causing application crashes

>> Comparing results with other developers

>> Creating a baseline test for custom data testing later

>> Verifying the adequacy of error-trapping code used for issues such as missing data

>> Ensuring that graphs and plots appear as they should

>> Saving time creating a test dataset

>> Devising mock-ups for demo purposes that don't compromise sensitive custom data

REMEMBER

A standardized common dataset is just a starting point, however. At some point, you need to verify that your own custom data works, but after verifying that the standard dataset works, you can do so with more confidence in the reliability of your application code. Perhaps the best reason to use one of these datasets is to reduce the time needed to locate and fix errors of various sorts — errors that might otherwise prove time consuming because you couldn't be sure of the data that you're using.

Finding the Right Dataset

Locating the right dataset for testing purposes is essential. Fortunately, you don't have to look very hard because some online sites provide you with everything needed to make a good decision. The following sections offer insights into locating the right dataset for your needs.

Locating general dataset information

Datasets appear in a number of places online, and you can use many of them for general needs. An example of these sorts of datasets appears on the UCI Machine Learning Repository at http://archive.ics.uci.edu/ml/datasets.html, shown in Figure 15-1. As the table shows, the site categorizes the individual datasets so that you can find the dataset you need. More important, the table helps you understand the kinds of tasks that people normally employ the dataset to perform.

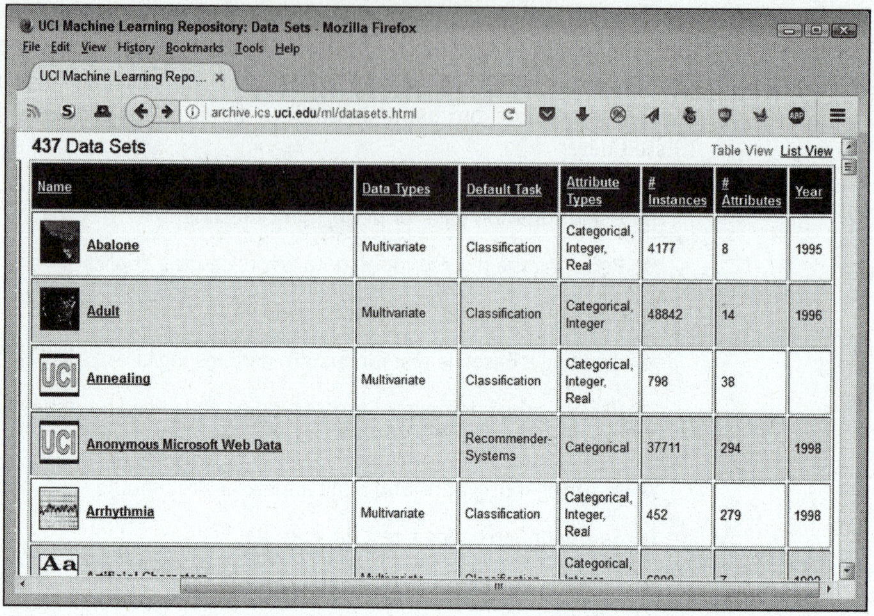

FIGURE 15-1: Standardized, common, datasets are categorized in specific ways.

UCI Machine Learning Repository: Data Sets - Mozilla Firefox

File Edit View History Bookmarks Tools Help

archive.ics.uci.edu/ml/datasets.html

437 Data Sets   Table View List View

Name	Data Types	Default Task	Attribute Types	# Instances	# Attributes	Year
Abalone	Multivariate	Classification	Categorical, Integer, Real	4177	8	1995
Adult	Multivariate	Classification	Categorical, Integer	48842	14	1996
Annealing	Multivariate	Classification	Categorical, Integer, Real	798	38	
Anonymous Microsoft Web Data		Recommender-Systems	Categorical	37711	294	1998
Arrhythmia	Multivariate	Classification	Categorical, Integer, Real	452	279	1998
Artificial Character...	Multivariate	Classification	Categorical, Integer	6000	7	1992

If you want to know more about a particular dataset, you click its link and go to a page like the one shown in Figure 15-2. You can determine whether a dataset will help you test certain application features, such as searching for and repairing missing values. The Number of Web Hits field tells you how popular the dataset is, which can affect your ability to find others who have used the dataset for testing purposes. All this information is helpful in ensuring that you get the right dataset for a particular need; the goals include error detection, performance testing, and comparison with other applications of the same type.

TIP

Even if your language provides easy access to these datasets, getting onto a site such as UCI Machine Learning Repository can help you understand which of these datasets will work best. In many cases, a language will provide access to the dataset and a brief description of dataset content — not a complete description of the sort you find on this site.

Using library-specific datasets

Depending on your programming language, you likely need to use a library to work with datasets in any meaningful way. One such library for Python is Scikit-learn (http://scikit-learn.org/stable/). This is one of the more popular libraries because it contains such an extensive set of features and also provides the means for loading both internal and external datasets as described at http://scikit-learn.org/stable/datasets/index.html. You can obtain various kinds of datasets using Scikit-learn as follows:

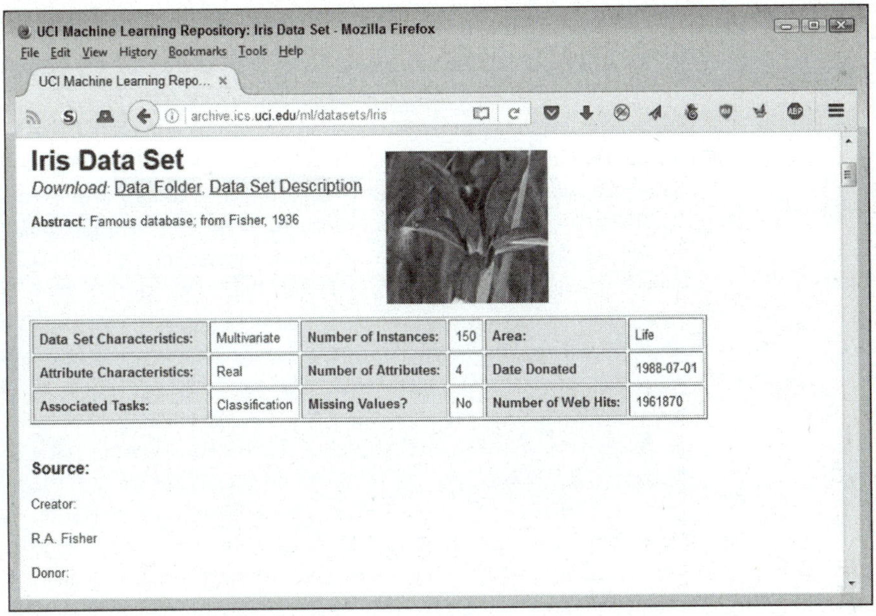

FIGURE 15-2:
Dataset details are important because they help you find the right dataset.

>> **Toy datasets:** Provides smaller datasets that you can use to test theories and basic coding.

>> **Image datasets:** Includes datasets containing basic picture information that you can use for various kinds of graphic analysis.

>> **Generators:** Defines randomly generated data based on the specifications you provide and the generator used. You can find generators for

- Classification and clustering

- Regression

- Manifold learning

- Decomposition

>> **Support Vector Machine (SVM) datasets:** Provides access to both the svmlight (`http://svmlight.joachims.org/`) and libsvm (`https://www.csie.ntu.edu.tw/~cjlin/libsvm/`) implementations, which include datasets that enable you to perform sparse dataset tasks.

>> **External load:** Obtains datasets from external sources. Python provides access to a huge number of datasets, each of which is useful for a particular kind of analysis or comparison. When accessing an external dataset, you may have to rely on additional libraries:

- `pandas.io`: Provides access to common data formats that include CSV, Excel, JSON, and SQL.

- `scipy.io`: Obtains information from binary formats popular with the scientific community, including `.mat` and `.arff` files.

- `numpy/routines.io`: Loads columnar data into NumPy (`http://www.numpy.org/`) arrays.

- `skimage.io`: Loads images and videos into NumPy arrays.

- `scipy.io.wavfile.read`: Reads `.wav` file data into NumPy arrays.

>> **Other:** Includes standard datasets that provide enough information for specific kinds of testing in a real-world manner. These datasets include (but are not limited to) Olivetti Faces and 20 Newsgroups Text.

Loading a Dataset

The fact that Python provides access to such a large variety of datasets might make you think that a common mechanism exists for loading them. Actually, you need a variety of techniques to load even common datasets. As the datasets become more esoteric, you need additional libraries and other techniques to get the job done. The following sections don't give you an exhaustive view of dataset loading in Python, but you do get a good overview of the process for commonly used datasets so that you can use these datasets within the functional programming environment. (See the "Finding Haskell support" sidebar in this chapter for reasons that Haskell isn't included in the sections that follow.)

Working with toy datasets

As previously mentioned, a toy dataset is one that contains a small amount of common data that you can use to test basic assumptions, functions, algorithms, and simple code. The toy datasets reside directly in Scikit-learn, so you don't have to do anything special except call a function to use them. The following list provides a quick overview of the function used to import each of the toy datasets into your Python code:

>> `load_boston()`: Regression analysis with the Boston house-prices dataset

>> `load_iris()`: Classification with the iris dataset

>> `load_diabetes()`: Regression with the diabetes dataset

>> `load_digits([n_class])`: Classification with the digits dataset

>> `load_linnerud()`: Multivariate regression using the linnerud dataset (health data described at https://github.com/scikit-learn/scikit-learn/blob/master/sklearn/datasets/descr/linnerud.rst)

>> `load_wine()`: Classification with the wine dataset

>> `load_breast_cancer()`: Classification with the Wisconsin breast cancer dataset

REMEMBER

Note that each of these functions begins with the word *load.* When you see this formulation in Python, the chances are good that the associated dataset is one of the Scikit-learn toy datasets.

The technique for loading each of these datasets is the same across examples. The following example shows how to load the Boston house-prices dataset:

```
from sklearn.datasets import load_boston
Boston = load_boston()
print(Boston.data.shape)
```

To see how the code works, click Run Cell. The output from the `print()` call is (506, 13). You can see the output shown in Figure 15-3.

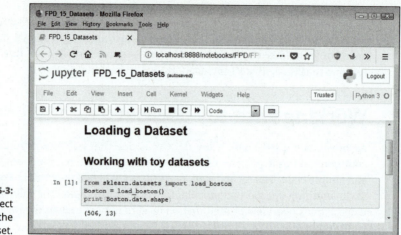

FIGURE 15-3:
The Boston object contains the loaded dataset.

Creating custom data

The purpose of each of the data generator functions is to create randomly generated datasets that have specific attributes. For example, you can control the number of data points using the n_samples argument and use the centers argument to control how many groups the function creates within the dataset. Each of the calls starts with the word *make*. The kind of data depends on the function; for example, make_blobs() creates Gaussian blobs for clustering (see http://scikit-learn.org/stable/modules/generated/sklearn.datasets.make_blobs.html for details). The various functions reflect the kind of labeling provided: single label and multilabel. You can also choose bi-clustering, which allows clustering of both matrix rows and columns. Here's an example of creating custom data:

```
from sklearn.datasets import make_blobs
X, Y = make_blobs(n_samples=120, n_features=2, centers=4)
print(X.shape)
```

The output will tell you that you have indeed created an X object containing a dataset with two features and 120 cases for each feature. The Y object contains the color values for the cases. Seeing the data plotted using the following code is more interesting:

```
import matplotlib.pyplot as plt
%matplotlib inline
plt.scatter(X[:, 0], X[:, 1], s=25, c=Y)
plt.show()
```

The %matplotlib magic function appears in Table 11-1. In this case, you tell Notebook to present the plot inline. The output is a scatter chart using the x-axis and y-axis contained in X. The c=Y argument tells scatter() to create the chart using the color values found in Y. Figure 15-4 shows the output of this example. Notice that you can clearly see the four clusters based on their color (even though the colors don't appear in the book).

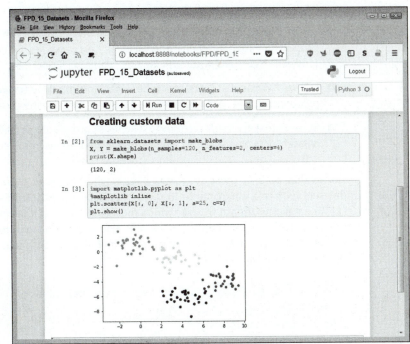

FIGURE 15-4: Custom datasets provide randomized data output in the form you specify.

Fetching common datasets

At some point, you need larger datasets of common data to use for testing. The toy datasets that worked fine when you were testing your functions may not do the job any longer. Python provides access to larger datasets that help you perform more complex testing but won't require you to rely on network sources. These datasets will still load on your system so that you're not waiting on network latency during testing. Consequently, they're between the toy datasets and a real-world dataset in size. More important, because they rely on actual (standardized) data, they reflect real-world complexity. The following list tells you about the common datasets:

» `fetch_olivetti_faces()`: Olivetti faces dataset from AT&T containing ten images each of 40 different test subjects; each grayscale image is 64 x 64 pixels in size

» `fetch_20newsgroups(subset='train')`: Data from 18,000 newsgroup posts based on 20 topics, with the dataset split into two subgroups: one for training and one for testing

» `fetch_mldata('MNIST original', data_home=custom_data_home)`: Dataset containing machine learning data in the form of 70,000, 28-x-28-pixel handwritten digits from 0 through 9

» `fetch_lfw_people(min_faces_per_person=70, resize=0.4)`: Labeled Faces in the Wild dataset described at `http://vis-www.cs.umass.edu/lfw/`, which contains pictures of famous people in JPEG format

» `sklearn.datasets.fetch_covtype()`: U.S. forestry dataset containing the predominant tree type in each of the patches of forest in the dataset

» `sklearn.datasets.fetch_rcv1()`: Reuters Corpus Volume I (RCV1) is a dataset containing 800,000 manually categorized stories from Reuters, Ltd.

Notice that each of these functions begins with the word *fetch*. Some of these datasets require a long time to load. For example, the Labeled Faces in the Wild (LFW) dataset is 200MB in size, which means that you wait several minutes just to load it. However, at 200MB, the dataset also begins (in small measure) to start reflecting the size of real-world datasets. The following code shows how to fetch the Olivetti faces dataset:

```
from sklearn.datasets import fetch_olivetti_faces
data = fetch_olivetti_faces()
print(data.images.shape)
```

When you run this code, you see that the shape is 400 images, each of which is 64 x 64 pixels. The resulting `data` object contains a number of properties, including images. To access a particular image, you use `data.images[?]`, where ? is the number of the image you want to access in the range from 0 to 399. Here is an example of how you can display an individual image from the dataset.

```
import matplotlib.pyplot as plt
%matplotlib inline
plt.imshow(data.images[1], cmap="gray")
plt.show()
```

TIP

The `cmap` argument tells how to display the image, which is in grayscale in this case. The tutorial at `https://matplotlib.org/tutorials/introductory/images.html` provides additional information on using `cmap`, as well as on adjusting the image in various ways. Figure 15-5 shows the output from this example.

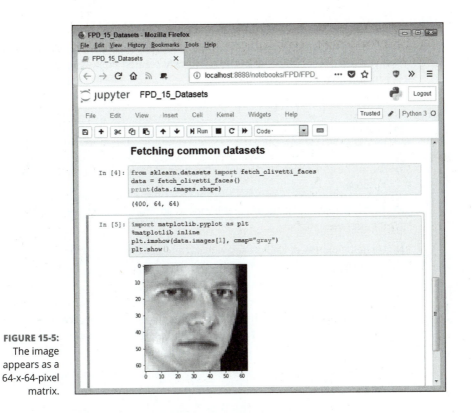

Manipulating Dataset Entries

You're unlikely to find a common dataset used with Python that doesn't provide relatively good documentation. You need to find the documentation online if you want the full story about how the dataset is put together, what purpose it serves, and who originated it, as well as any needed statistics. Fortunately, you can employ a few tricks to interact with a dataset without resorting to major online research. The following sections offer some tips for working with the dataset entries found in this chapter.

Determining the dataset content

The previous sections of this chapter show how to load or fetch existing datasets from specific sources. These datasets generally have specific characteristics that you can discover online at places like http://scikit-learn.org/stable/modules/generated/sklearn.datasets.load_boston.html for the Boston house-prices dataset. However, you can also use the dir() function to learn about dataset content. When you use dir(Boston) with the previously created Boston house-prices dataset, you discover that it contains DESCR, data, feature_names, and target properties. Here is a short description of each property:

» DESCR: Text that describes the dataset content and some of the information you need to use it effectively

» data: The content of the dataset in the form of values used for analysis purposes

» feature_names: The names of the various attributes in the order in which they appear in data

» target: An array of values used with data to perform various kinds of analysis

The print(Boston.DESCR) function displays a wealth of information about the Boston house-prices dataset, including the names of attributes that you can use to interact with the data. Figure 15-6 shows the results of these queries.

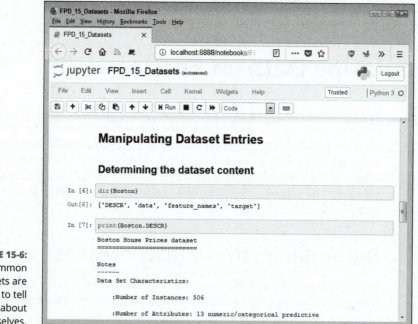

FIGURE 15-6: Most common datasets are configured to tell you about themselves.

REMEMBER

The information that the datasets contain can have significant commonality. For example, if you use `dir(data)` for the Olivetti faces dataset example described earlier, you find that it provides access to `DESCR`, `data`, `images`, and `target` properties. As with the Boston house-prices dataset, `DESCR` gives you a description of the Olivetti faces dataset, which you can use for things like accessing particular attributes. By knowing the names of common properties and understanding how to use them, you can discover all you need to know about a common dataset in most cases without resorting to any online resource. In this case, you'd use `print(data.DESCR)` to obtain a description of the Olivetti faces dataset. Also, some of the description data contains links to sites where you can learn more information.

Creating a DataFrame

The common datasets are in a form that allows various types of analysis, as shown by the examples provided on the sites that describe them. However, you might not want to work with the dataset in that manner; instead, you may want something that looks a bit more like a database table. Fortunately, you can use the pandas (`https://pandas.pydata.org/`) library to perform the conversion in a manner that makes using the datasets in other ways easy. Using the Boston house-prices dataset as an example, the following code performs the required conversion:

```
import pandas as pd
BostonTable = pd.DataFrame(Boston.data,
                    columns=Boston.feature_names)
```

If you want to include the target values with the DataFrame, you must also execute: BostonTable['target'] = Boston.target. However, this chapter doesn't use the target data.

Accessing specific records

If you were to do a dir() command against a DataFrame, you would find that it provides you with an overwhelming number of functions to try. The documentation at https://pandas.pydata.org/pandas-docs/version/0.23/generated/pandas.DataFrame.html supplies a good overview of what's possible (which includes all the usual database-specific tasks specified by CRUD). The following example code shows how to perform a query against a pandas DataFrame. In this case, the code selects only those housing areas where the crime rate is below 0.02 per capita.

```
CRIMTable = BostonTable.query('CRIM < 0.02')
print(CRIMTable.count()['CRIM'])
```

The output shows that only 17 records match the criteria. The count() function enables the application to count the records in the resulting CRIMTable. The index, ['CRIM'], selects just one of the available attributes (because every column is likely to have the same values).

You can display all these records with all of the attributes, but you may want to see only the number of rooms and the average house age for the affected areas. The following code shows how to display just the attributes you actually need:

```
print(CRIMTable[['RM', 'AGE']])
```

Figure 15-7 shows the output from this code. As you can see, the houses vary between 5 and nearly 8 rooms in size. The age varies from almost 14 years to a little over 65 years.

You might find it a bit hard to work with the unsorted data in Figure 15-7. Fortunately, you do have access to the full range of common database features. If you want to sort the values by number of rooms, you use:

```
print(CRIMTable[['RM', 'AGE']].sort_values('RM'))
```

As an alternative, you can always choose to sort by average home age:

```
print(CRIMTable[['RM', 'AGE']].sort_values('AGE'))
```

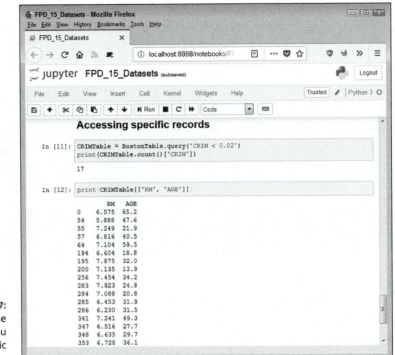

FIGURE 15-7:
Manipulating the data helps you find specific information.

5
Performing Simple Error Trapping

Chapter **16**

Handling Errors in Haskell

Most application code contains errors. It's a blanket statement that you may doubt, but the wealth of errors is obvious when you consider the number of security breaches and hacks that appear in the trade press, not to mention the odd results that sometimes occur from seemingly correct data analysis. If the code has no bugs, updates will occur less often. This chapter discusses errors from a pure functional language perspective; Chapter 17 looks at the same issue from an impure language perspective, which can differ because impure languages often rely on procedures.

After you identify an error, you can describe the error in detail and use that description to locate the error in the application code. At least, this process is the theory that most people go by when finding errors. Reality is different. Errors commonly hide in plain view because the developer isn't squinting just the right way in order to see them. Bias, perspective, and lack of understanding all play a role in hiding errors from view. This chapter also describes how to locate and describe errors so that they become easier to deal with.

Knowing the source, location, and complete description of an error doesn't fix the error. People want applications that provide a desired result based on specific inputs. If your application doesn't provide this sort of service, people will stop using it. To keep people from discarding your application, you need to correct the error or handle the situation that creates the environment in which the error occurs. The final section of this chapter describes how to squash errors —for most of the time, at least.

Defining a Bug in Haskell

A *bug* occurs when an application either fails to run or produces an output other than the one expected. An infinite loop is an example of the first bug type, and obtaining a result of 5 when adding 1 and 1 is an example of the second bug type. Some people may try to convince you that other kinds of bugs exist, but these other bugs end up being subsets of the two just mentioned.

WARNING

Haskell and other functional languages don't allow you to write applications that are bug free. Quite the contrary: You can find the same sorts of bugs in Haskell that you can find in other languages, such as Python. Chapter 17 explores some common Python issues and examines the conditions under which bugs occur in that language, but many of those issues also translate into Haskell. Bugs occur at compile time or runtime. In addition, they can be syntactical, semantic, or logical in nature.

However, functional languages tend to bring their own assortment of bugs into applications, and knowing what these bugs are is a good idea. They're not necessarily new bugs, but they occur differently with functional languages. The following sections consider the specifics of bugs that occur with functional languages, using Haskell as an example. These sections provide an overview of the kinds of Haskell-specific bugs that you need to think about, but you can likely find others.

Considering recursion

Functional languages generally avoid mutable variables by using recursion. This difference in focus means that you're less apt to see logic errors that occur when loops don't execute the number of times expected or fail to stop because the condition that you expected doesn't occur. However, it also means that stack-related errors from infinite recursion happen more often.

WARNING

You may think that loops and recursion produce similar errors. However, unlike a loop, recursion can't go on indefinitely because the stack uses memory for each call, which means that the application eventually runs out of memory. In fact, memory helps define the difference between functional and other languages that do rely on loops. When a functional language runs out of memory to perform recursion, the problem could simply be that the host machine lacks the required resources, rather than an actual code error.

INTRODUCING THE ALGORITHM CONNECTION

According to the National Public Radio (NPR) article at https://www.npr.org/sections/alltechconsidered/2015/03/23/394827451/now-algorithms-are-deciding-whom-to-hire-based-on-voice, an algorithm can decide whether a company hires you for a job based solely on your voice. The algorithm won't make the final decision, but it does reduce the size of the list that a human will go through to make the final determination. If a human never sees your name, you'll never get the job. The problem is that algorithms contain a human element.

The Big Think article at https://bigthink.com/ideafeed/when-algorithms-go-awry discusses the issue of human thought behind the algorithm. The laws that define human understanding of the universe today rely on the information at hand, which constantly changes. Therefore, the laws constantly change as well. Given that the laws are, at best, unstable and that functional languages rely heavily on algorithms presented in a specific manner, the bug you're hunting may have nothing to do with your code; it may instead have everything to do with the algorithm you're using. You can find more information about the biases and other issues surrounding algorithms in *Algorithms For Dummies,* by John Mueller and Luca Massaron (Wiley).

The problem with algorithms goes deeper, however, than simply serving as the basis on which someone creates the algorithm. Absolute laws tend to pervert the intent of a particular set of rules. Unlike humans, computers execute only the instructions that humans provide; a computer can't understand the concept of exceptions. In addition, no one can provide a list of every exception as part of an application. Consequently, algorithms, no matter how well constructed, will eventually become incorrect because of changes in the information used to create them and the incapability of an algorithm to adapt to exceptions.

Understanding laziness

Haskell is a *lazy language* for the most part, which means that it doesn't perform actions until it actually needs to perform them. For example, it won't evaluate an expression until it needs to use the output from that expression. The advantages of using a lazy language include (but aren't limited to) the following:

>> Faster execution speed because an expression doesn't use processing cycles until needed

>> Reduced errors because an error shows up only when the expression is evaluated

- >> Reduced resource usage because resources are used only when needed

- >> Enhanced ability to create data structures that other languages can't support (such as a data structure of infinite size)

- >> Improved control flow because you can define some objects as abstractions rather than primitives

REMEMBER

However, lazy languages can also create strange bug scenarios. For example, the following code purports to open a file and then read its content:

```
withFile "MyData.txt" ReadMode handle >>= putStr
```

If you looked at the code from a procedural perspective, you would think that it should work. The problem is that lazy evaluation using `withFile` means that Haskell closes `handle` before it reads the data from `MyData.txt`. The solution to the problem is to perform the task as part of a do, like this:

```
main = withFile "MyData.txt" ReadMode $ \handle -> do
          myData <- hGetLine handle
          putStrLn myData
```

However, by the time you create the code like this, it really isn't much different from the example found in the "Reading data" section of Chapter 13. The main advantage is that Haskell automatically closes the file handle for you. Offsetting this advantage is that the example in Chapter 13 is easier to read. Consequently, lazy evaluation can impose certain unexpected restrictions.

Using unsafe functions

Haskell generally provides safe means of performing tasks, as mentioned in several previous chapters. Not only is type safety ensured, but Haskell also checks for issues such as the correct number of inputs and even the correct usage of outputs. However, you may encounter extremely rare circumstances in which you need to perform tasks in an unsafe manner in Haskell, which means using unsafe functions of the sort described at `https://wiki.haskell.org/Unsafe_functions`. Most of these functions are fully described as part of the `System.IO.Unsafe` package at `http://hackage.haskell.org/package/base-4.11.1.0/docs/System-IO-Unsafe.html`. The problem is that these functions are, as described, unsafe and therefore the source of bugs in many cases.

REMEMBER

You can find the rare exceptions for using unsafe functions in posts online. For example, you might want to access the functions in the C math library (as accessed through `math.h`). The discussion at `https://stackoverflow.com/questions/10529284/is-there-ever-a-good-reason-to-use-unsafeperformio` tells how to perform this task. However, you need to consider whether such access is really needed because Haskell provides such an extensive array of math functions.

The same discussion explores other uses for unsafePerformIO. For example, one of the code samples shows how to create global mutable variables in Haskell, which would seem counterproductive, given the reason you're using Haskell in the first place. Avoiding unsafe functions in the first place is a better idea because you open yourself to hours of debugging, unassisted by Haskell's built-in functionality (after all, you marked the call as unsafe).

Considering implementation-specific issues

As with most language implementations, you can experience implementation-specific issues with Haskell. This book uses the Glasgow Haskell Compiler (GHC) version 8.2.2, which comes with its own set of incompatibilities as described at `http://downloads.haskell.org/~ghc/8.2.2/docs/html/users_guide/bugs.html`. Many of these issues will introduce subtle bugs into your code, so you need to be aware of them. When you run your code on other systems using other implementations, you may find that you need to rework the code to bring it into compliance with that implementation, which may not necessarily match the Haskell standard.

Understanding the Haskell-Related Errors

It's essential to understand that the functional nature of Haskell and its use of expressions modifies how people commonly think about errors. For example, if you type **x = 5/0** and press Enter in Python, you see a ZeroDivisionError as output. In fact, you expect to see this sort of error in any procedural language. On the other hand, if you type **x = 5/0** in Haskell and press Enter, nothing seems to happen. However, x now has the value of Infinity. The fact that some pieces of code that define an error in a procedural language but may not define an error in a functional language means that you need to be aware of the consequences.

To see the consequences in this case, type **:t x** and press Enter. You find that the type of x is Fractional, not Float or Double as you might suppose. Actually, you can convert x to either Float or Double by typing **y = x::Double** or **y = x::Float** and pressing Enter.

The Fractional type is a superset of both Double and Float, which can lead to some interesting errors that you don't find in other languages. Consider the following code:

```
x = 5/2
:t x
y = (5/2)::Float
:t y
z = (5/2)::Double
:t z
x * y
:t (x * y)
x * z
:t (x * z)
y * z
```

The code assigns the same values to three variables, x, y, and z, but of different types: Fractional, Float, and Double. You verify this information using the :t command. The first two multiplications work as expected and produce the type of the subtype, rather than the host, Fractional. However, notice that trying to multiply a Float by a Double, something you could easily do in most procedural languages, doesn't work in Haskell, as shown in Figure 16-1. You can read about the reason for the lack of automatic type conversion in Haskell at https://wiki.haskell.org/Generic_number_type. To make this last multiplication work, you need to convert one of the two variables to Fractional first using code like this: realToFrac(y) * z.

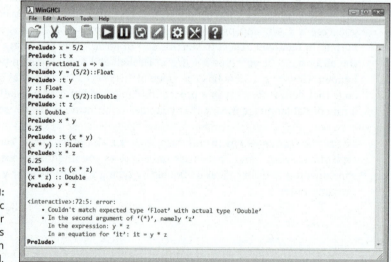

FIGURE 16-1: Automatic number conversion is unavailable in Haskell.

REDUCING THE NUMBER OF BUGS

Some people will try to convince you that one language or another provides some sort of magic that reduces bugs to nearly zero without any real effort on your part. Unfortunately, those people are wrong. In fact, you could say that accurately comparing languages against each other for bug deterrence is impossible. Someone who is skilled in one language but not another will almost certainly produce more bugs in the latter, despite any protections that the latter has. In addition, any developer is unlikely to have precisely the same level of skill in using two languages, so the comparison isn't meaningful. Consequently, the language that produces the fewest bugs is often the language you know best.

Another important issue to consider is that programmers tend to use differing meanings for the word *bug*, which is why this chapter attempts to provide a fairly comprehensive definition — albeit one that you may not agree with. If the target for analysis isn't fully defined, you can't perform any meaningful comparison. Before you could hope to determine which language produces the fewest bugs, you would need to have agreement on what constitutes a bug, and such agreement doesn't currently exist.

Even more important is the concept of what a bug means to nondevelopers. A developer will look for correct output for specific input. However, a user may see a bug in presenting three digits past the decimal point, instead of just two. A manager may see a bug in presenting output that doesn't match company policies and should be vetted to ensure that the output is consistent with those polices. An administrator may see a bug in a suggested fix for an error message that runs counter to security requirements. Consequently, you must also perform stakeholder testing, and adding this level of testing makes it even harder to compare languages, environments, testing methodologies, and a whole host of other concerns that affect that seemingly simple word *bug*. So if you were hoping to find some meaningful comparison between the relative numbers of bugs that Haskell produces versus those created by Python in this book, you'll be disappointed.

REMEMBER

Some odd situations exist in which a Haskell application can enter an infinite loop because it works with expressions rather than relying on procedures. For example, the following code will execute fine in Python:

```
x = 5/2
x = x + 1
x
```

In Python, you see an output of 3.5, which is what anyone working with procedural code will expect. However, this same code causes Haskell to enter into an infinite loop because the information is evaluated as an expression, not as a procedure. The output, when working with compiled code, is `<<loop>>`, which you can read about in more detail `https://stackoverflow.com/questions/21505192/haskell-program-outputs-loop`. When using WinGHCi (or another interpreter), the call will simply never return. You need to click the Pause button (which looks like the Pause button on a remote) instead. A message of `Interrupted` appears to tell you that the code, which will never finish its work, has been interrupted. The fact that Haskell actually detects many simpler infinite loops and tells you about them says a lot about its design.

TIP

Haskell does prevent a wide variety of errors that you see in many other languages. For example, it doesn't have a global state. Therefore, one function can't use a global variable to corrupt another function. The type system also prevents a broad range of errors that plague other languages, such as trying to stuff too much data into a variable that can't hold it. You can read a discussion of other sorts of common errors that Haskell prevents at `https://www.quora.com/Exactly-what-kind-of-bugs-does-Haskell-prevent-from-introducing-compared-to-other-mainstream-languages`.

Even though this section isn't a complete list of all the potential kinds of errors that you see in Haskell, understand that functional languages have many similarities in the potential sources of errors but that the actual kinds of errors can differ.

Fixing Haskell Errors Quickly

Haskell, as you've seen in the error messages in this book, is good about providing you with trace information when it does encounter an error. Errors can occur in a number of ways, as described in Chapter 17. Of course, the previous sections have filled you in on Haskell exceptions to the general rules. The following sections give an overview of some of the ways to fix Haskell errors quickly.

Relying on standard debugging

Haskell provides the usual number of debugging tricks, and the IDE you use may provide others. Because of how Haskell works, your first line of defense against bugs is in the form of the messages, such as error and CallStack output, that Haskell provides. Figure 16-1 shows an example of an error output, and Figure 16-2 shows an example of CallStack output. Comparing the two, you can see that they're quite similar. The point is that you can use this output to trace the origin of a bug in your code.

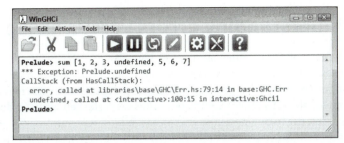

FIGURE 16-2:
Haskell provides
you with
reasonably
useful messages
in most cases.

During the debugging process, you can use the trace function to validate your assumptions. To use trace, you must import Debug.Trace. Figure 16-3 shows a quick example of this function at work.

FIGURE 16-3:
Use trace to
validate your
assumptions.

REMEMBER

You provide the assumption as a string in the first argument and the function call as the second argument. The article at http://hackage.haskell.org/package/base-4.11.1.0/docs/Debug-Trace.html gives additional details on using trace. Note that with lazy execution, you see trace output only when Haskell actually executes your code. Consequently, in contrast to other development languages, you may not see all your trace statements every time you run the application. A specialized alternative to trace is htrace, which you can read about at http://hackage.haskell.org/package/htrace.

Haskell does provide other debugging functionality. For example, you gain full access to breakpoints. As with other languages, you have methods available for determining the status of variables when your code reaches a breakpoint (assuming that the breakpoint actually occurs with lazy execution). The article at https://wiki.haskell.org/Debugging offers additional details.

Understanding errors versus exceptions

For most programming languages, you can use the terms *error* and *exception* almost interchangeably because they both occur for about the same reasons. Some languages purport to provide a different perspective on the two but then fail to support the differences completely. However, Haskell actually does differentiate between the two:

>> **Error:** An error always occurs as the result of a mistake in the code. The error is never expected and you must fix it to make the code run properly. The functions that support errors are

- `error`
- `assert`
- `Control.Exception.catch`
- `Debug.Trace.trace`

>> **Exception:** An exception is an expected, but unusual, occurrence. In many cases, exceptions reflect conditions outside the application, such as a lack of drive space or an incapability to create a connection. You may not be able to fix an exception but you can sometimes compensate for it. The function that support exceptions are

- `Prelude.catch`
- `Control.Exception.catch`
- `Control.Exception.try`
- `IOError`
- `Control.Monad.Error`

REMEMBER

As you can see, errors and exceptions fulfill completely different purposes and generally use different functions. The only repeat is `Control.Exception.catch`, and there are some caveats about using this function for an error versus an exception, as described at `https://wiki.haskell.org/Error_vs._Exception`. This article also gives you additional details about the precise differences between errors and exceptions.

Chapter **17**

Handling Errors in Python

C hapter 16 discusses errors in code from a Haskell perspective, and some of the errors you encounter in Haskell might take you by surprise. Oddly enough, so might some of the coding techniques used in other languages that would appear as errors. (Chapter 16 also provides a good reason not to compare the bug mitigation properties of various languages in the "Reducing the number of bugs" sidebar.) Python is more traditional in its approach to errors. For example, dividing a number by zero actually does produce an error, not a special data type designed to handle the division using the value Infinity. Consequently, you may find the discussion (in the first section of this chapter) of what constitutes a bug in Python a little boring if you have worked through coding errors in other procedural languages. Even so, reading the material is a good idea so that you can better understand how Python and Haskell differ in their handling of errors in the functional programming environment.

The next section of the chapter goes into the specifics of Python-related errors, especially those related to the functional features that Python provides. Although the chapter does contain a little general information as background, it focuses mostly on the functional programming errors.

Finally, the chapter tells you about techniques that you can use to fix Python functional programming errors a little faster. You'll find the same sorts of things that you can do when using Python for procedural programming, such as step-by-step debugging. However, fixing functional errors sometimes requires a different thought process, and this chapter helps you understand what you need to do when such cases arise.

Defining a Bug in Python

As with Haskell, Python bugs occur when an application fails to work as anticipated. Both languages also view errors that create bugs in essentially the same manner, even though Haskell errors take a functional paradigm's approach, while those in Python are more procedural in nature. The following sections help you understand what is meant by a bug in Python and provide input on how using the functional approach can affect the normal view of bugs.

Considering the sources of errors

You might be able to divine the potential sources of error in your application by reading tea leaves, but that's hardly an efficient way to do things. Errors actually fall into well-defined categories that help you predict (to some degree) when and where they'll occur. By thinking about these categories as you work through your application, you're far more likely to discover potential errors' sources before they occur and cause potential damage. The two principal categories are

» Errors that occur at a specific time

» Errors that are of a specific type

The following sections discuss these two categories in greater detail. The overall concept is that you need to think about error classifications in order to start finding and fixing potential errors in your application before they become a problem.

Classifying when errors occur

Errors occur at specific times. However, no matter when an error occurs, it causes your application to misbehave. The two major time frames in which errors occur are

» **Compile time:** A compile time error occurs when you ask Python to run the application. Before Python can run the application, it must interpret the code and put it into a form that the computer can understand. A computer relies on machine code that is specific to that processor and architecture. If the instructions you write are malformed or lack needed information, Python can't perform the required conversion. It presents an error that you must fix before the application can run.

» **Runtime:** A runtime error occurs after Python compiles the code that you write and the computer begins to execute it. Runtime errors come in several different types, and some are harder to find than others. You know you have a runtime error when the application suddenly stops running and displays an exception dialog box or when the user complains about erroneous output (or at least instability).

REMEMBER

Not all runtime errors produce an exception. Some runtime errors cause instability (the application freezes), errant output, or data damage. Runtime errors can affect other applications or create unforeseen damage to the platform on which the application is running. In short, runtime errors can cause you quite a bit of grief, depending on precisely the kind of error you're dealing with at the time.

Distinguishing error types

You can distinguish errors by type, that is, by how they're made. Knowing the error types helps you understand where to look in an application for potential problems. Exceptions work like many other things in life. For example, you know that electronic devices don't work without power. So when you try to turn your television on and it doesn't do anything, you might look to ensure that the power cord is firmly seated in the socket.

TIP

Understanding the error types helps you locate errors faster, earlier, and more consistently, resulting in fewer misdiagnoses. The best developers know that fixing errors while an application is in development is always easier than fixing it when the application is in production because users are inherently impatient and want errors fixed immediately and correctly. In addition, fixing an error earlier in the development cycle is always easier than fixing it when the application nears completion because less code exists to review.

The trick is to know where to look. With this in mind, Python (and most other programming languages) breaks errors into the following types (arranged in order of difficulty, starting with the easiest to find):

» **Syntactical:** Whenever you make a typo of some sort, you create a syntactical error. Some Python syntactical errors are quite easy to find because the application simply doesn't run. The interpreter may even point out the error for you by highlighting the errant code and displaying an error message.

However, some syntactical errors are quite hard to find. Python is case sensitive, so you may use the wrong case for a variable in one place and find that the variable isn't quite working as you thought it would. Finding the one place where you used the wrong capitalization can be quite challenging.

>> **Semantic:** When you create a loop that executes one too many times, you don't generally receive any sort of error information from the application. The application will happily run because it thinks that it's doing everything correctly, but that one additional loop can cause all sorts of data errors. When you create an error of this sort in your code, it's called a *semantic error*. Semantic errors are tough to find, and you sometimes need some sort of debugger to find them.

>> **Logical:** Some developers don't create a division between semantic and logical errors, but they are different. A semantic error occurs when the code is essentially correct but the implementation is wrong (such as having a loop execute once too often). Logical errors occur when the developer's thinking is faulty. In many cases, this sort of error happens when the developer uses a relational or logical operator incorrectly. However, logical errors can happen in all sorts of other ways, too. For example, a developer might think that data is always stored on the local hard drive, which means that the application may behave in an unusual manner when it attempts to load data from a network drive instead. Logical errors are quite hard to fix because the problem isn't with the actual code, yet the code itself is incorrectly defined. The thought process that went into creating the code is faulty; therefore, the developer who created the error is less likely to find it. Smart developers use a second pair of eyes to help spot logical errors.

Considering version differences

Python is one of the few languages around today that has active support for two major language versions. Even though Python 2.*x* support will officially end in 2020 (see `https://pythonclock.org/` for details), you can bet that many developers will continue to use it until they're certain that the libraries they need come in a fully compatible Python 3.*x* form. However, the problem isn't just with libraries but also with processes, documentation, existing code, and all sorts of other things that could affect someone who is using functional programming techniques in Python.

REMEMBER

Although the Python community has worked hard to make the transition easier, you can see significant functional programming differences by reviewing the Python 2.*x* material at `https://docs.python.org/2/howto/functional.html` and comparing it to the Python 3.*x* material at `https://docs.python.org/3/howto/functional.html`. The transition will introduce bugs into your applications, some of them quite hard to find and others that the compiler will let you know about. Articles, such as the one at `http://sebastianraschka.com/Articles/2014_python_2_3_key_diff.html` can help you locate and potentially fix these issues. (Note especially the integer division differences stated by the article because they really can throw your functional code off in a manner that is particularly hard to find.)

Understanding the Python-Related Errors

You can encounter more than a few kinds of errors when working with Python code. This chapter doesn't provide exhaustive treatment of those errors. However, the following sections do offer some clues as to what might be wrong with your functional code, especially as it deals with lambda expressions.

Dealing with late binding closures

You need to realize that Python is late binding, which means that Python looks up the values of variables when it calls an inner function that is part of a loop only when the loop is completed. Consequently, rather than use individual values within a loop, what you see is the final value. For a demonstration of this issue, consider the following code:

```
def create_values(numValues):
    return [lambda x : i * x for i in range(numValues)]

for indValue in create_values(5):
    print(indValue(2))
```

This code creates the specified number of functions, one for each value in range(numValues), which is create_values(5) (five) in this case. The idea is to create an output of five values using a particular multiplier (which is indValue(2) in this case). You might assume that the first function call will be 0 (the value of i) * 2 (the value of x supplied as an input). However, the first function is never called while i is equal to 0. In fact, it gets called the first time only when its value is 4 — at the end of the loop. As a result, the output you see when you call this function is a series of 8s. To fix this code, you need to use the following create_values() code instead:

```
def create_values(numValues):
    return [lambda x, i=i : i * x for i in
            range(numValues)]
```

REMEMBER

This version of the code uses a trick to force the value of i to reflect the actual value produced by each of the values output by range(numValues). Instead of being part of the inner function, i is now provided as an input. You call the function in the same manner as before, but now the output is correct. Oddly enough, this particular problem isn't specific to lambda expressions; it can happen in any Python code. However, developers see it more often in this situation because the tendency is to use a lambda expression in this case.

TIP

You can find another example of this late-binding closure issue in the posting at https://bugs.python.org/issue27738 (with another fix like the one shown in this section). The discussion at https://stackoverflow.com/questions/1107210/python-lambda-problems provides another solution to this problem using functools.partial(). The point is that you must remember that Python is late binding.

Using a variable

In some situations, you can't use a lambda expression inline. Fortunately, Python will generally find these errors and tell you about them, as in the following code:

```
garbled = "IXX aXXmX sXeXcXrXeXt mXXeXsXsXaXXXXXXgXeX!XX"
print filter(lambda x: x != "X", garbled)
```

Obviously, this example is incredible simple, and you likely wouldn't use it in the real world. However, it shows that you can't use the lambda inline in this case; you must first assign it to a variable and then loop through the values. The following code shows the correct alternative code:

```
garbled = "IXX aXXmX sXeXcXrXeXt mXXeXsXsXaXXXXXXgXeX!XX"
ungarble = filter(lambda x: x != "X", garbled)
for x in ungarble:
    print(x, end='')
```

Working with third-party libraries

Your Python functional programming experience will include third-party libraries that may not always benefit from the functional programming approach. Before you assume that a particular approach will work, you should review potential sources of error online. For example, the following message thread discusses potential problems with using lambda expressions to perform an aggregation with Pandas: https://github.com/pandas-dev/pandas/issues/7186. In many cases, the community of developers will have alternatives for you to try, as happened in this case.

Fixing Python Errors Quickly

The key to fixing Python errors quickly is to have a strategy for dealing with each sort of error described in the "Distinguishing error types" section, earlier in this chapter. If Python doesn't recognize an error during the compilation process, it often generates an exception or you see unwanted behavior. The use of lambda expressions to define an application that relies on the functional paradigm doesn't really change things, but the use of lambda expressions can create special circumstances, such as those described in the "Introducing the algorithm connection" sidebar of Chapter 16. The following sections describe the mix of error-correction processes that you can employ when using Python in functional mode.

Understanding the built-in exceptions

Python comes with a host of built-in exceptions — far more than you might think possible. You can see a list of these exceptions at `https://docs.python.org/3.6/library/exceptions.html`. The documentation breaks the exception list down into categories. Here is a brief overview of the Python exception categories that you work with regularly:

>> **Base classes:** The base classes provide the essential building blocks (such as the `Exception` exception) for other exceptions. However, you might actually see some of these exceptions, such as the `ArithmeticError` exception, when working with an application.

>> **Concrete exceptions:** Applications can experience hard errors — errors that are hard to overcome because no good way to handle them exists or they signal an event that the application must handle. For example, when a system runs out of memory, Python generates a `MemoryError` exception. Recovering from this error is hard because it releasing memory from other uses isn't always possible. When the user presses an interrupt key (such as Ctrl+C or Delete), Python generates a `KeyboardInterrupt` exception. The application must handle this exception before proceeding with any other tasks.

>> **OS exceptions:** The operating system can generate errors that Python then passes along to your application. For example, if your application tries to open a file that doesn't exist, the operating system generates a `FileNotFoundError` exception.

>> **Warnings:** Python tries to warn you about unexpected events or actions that could result in errors later. For example, if you try to inappropriately use a resource, such as an icon, Python generates a `ResourceWarning` exception. You want to remember that this particular category is a warning and not an actual error: Ignoring it can cause you woe later, but you can ignore it.

Obtaining a list of exception arguments

The list of arguments supplied with exceptions varies by exception and by what the sender provides. You can't always easily figure out what you can hope to obtain in the way of additional information. One way to handle the problem is to simply print everything by using code like this:

```python
import sys
try:
    File = open('myfile.txt')
except IOError as e:
    for Arg in e.args:
        print(Arg)
else:
    print("File opened as expected.")
    File.close();
```

The `args` property always contains a list of the exception arguments in string format. You can use a simple `for` loop to print each of the arguments. The only problem with this approach is that you're missing the argument names, so you know the output information (which is obvious in this case), but you don't know what to call it.

TECHNICAL STUFF

A more complex method of dealing with the issue is to print both the names and the contents of the arguments. The following code displays both the names and the values of each of the arguments:

```python
import sys
try:
    File = open('myfile.txt')
except IOError as e:
    for Entry in dir(e):
        if (not Entry.startswith("_")):
            try:
                print(Entry, " = ", e.__getattribute__(Entry))
            except AttributeError:
                print("Attribute ", Entry, " not accessible.")
else:
    print("File opened as expected.")
    File.close();
```

In this case, you begin by getting a listing of the attributes associated with the error argument object using the dir() function. The output of the dir() function is a list of strings containing the names of the attributes that you can print. Only those arguments that don't start with an underscore (_) contain useful information about the exception. However, even some of those entries are inaccessible, so you must encase the output code in a second try...except block.

The attribute name is easy because it's contained in Entry. To obtain the value associated with that attribute, you must use the __getattribute() function and supply the name of the attribute you want. When you run this code, you see both the name and the value of each of the attributes supplied with a particular error argument object. In this case, the actual output is as follows:

```
args  =  (2, 'No such file or directory')
Attribute  characters_written  not accessible.
errno  =  2
filename  =  myfile.txt
filename2 = None
strerror  =  No such file or directory
winerror  =  None
with_traceback  =  <built-in method with_traceback of
    FileNotFoundError object at 0x0000000003416DC8>
```

Considering functional style exception handling

The previous sections of this chapter have discussed using exceptions, but as presented in previous chapters, Haskell actually discourages the use of exceptions, partly because they're indicative of state, and many functional programming aficionados discourage this use as well. The fact that Haskell does present exceptions as needed is proof that they're not absolutely forbidden, which is a good thing considering that in some situations, you really do need to use exceptions when working with Python.

However, when working in a functional programming environment with Python, you have some alternatives to using exceptions that are more in line with the functional programming paradigm. For example, instead of raising an exception as the result of certain events, you could always use a base value, as discussed at https://softwareengineering.stackexchange.com/questions/334769/functional-style-exception-handling.

Haskell also offers some specialized numeric handling that you might also want to incorporate as part of using Python. For example, as shown in Chapter 16, the `Fractional` type allows statements such as 5 / 0 in Haskell. The same statement produces an error in Python. Fortunately, you have access to the `fractions` package in Python, as described at https://docs.python.org/3/library/fractions.html.

TIP

Although the `fractions` package addresses some issues and you get a full fractional type, that package doesn't address the 5 / 0 problem; you still get a `ZeroDivisionError` exception. To avoid this final issue, you can use specialized techniques such as those found in the message thread at https://stackover flow.com/questions/27317517/make-division-by-zero-equal-to-zero. The point is that you have ways around exceptions in some cases if you want to use a more functional style of reporting. If you really want some of the advantages of using Haskell in your Python application, the hyphen module at https://github.com/tbarnetlamb/hyphen makes it possible.

6

The Part of Tens

Discover must-have Haskell libraries.

Discover must-have Python packages.

Gain employment using functional programming techniques.

Chapter **18**

Ten Must-Have Haskell Libraries

Haskell supports a broad range of libraries, which is why it's such a good product to use. Even though this chapter explores a few of the more interesting Haskell library offerings, you should also check out the rather lengthy list of available libraries at http://hackage.haskell.org/packages/. Chances are that you'll find a library to meet almost any need in that list.

REMEMBER

The problem is figuring out precisely which library to use and, unfortunately, the Hackage site doesn't really help much. The associated short descriptions are generally enough to get you pointed in the right direction, but experimentation is the only real way to determine whether a library will meet your needs. In addition, you should seek online reviews of the various libraries before you begin using them. Of course, that's part of the pleasure of development: discovering new tools to meet specific needs and then testing them yourself.

binary

To store certain kinds of data, you must be able to serialize it — that is, change it into a format that you can store on disk or transfer over a network to another machine. Serialization takes complex data structures and data objects and turns them into a series of bits that an application can later reconstitute into the original structure or object using deserialization. The point is that the data can't travel in its original form. The binary library (`http://hackage.haskell.org/package/binary`) enables an application to serialize binary data of the sort used for all sorts of purposes, including both sound and graphics files. It works on lazy byte strings, which can provide a performance advantage as long as the byte strings are error free and the code is well behaved.

TIP

This particular library's fast speed is why it's so helpful for real-time binary data needs. According to the originator, you can perform serialization and deserialization tasks at speeds approaching 1 Gbps. According to the discussion at `https://superuser.com/questions/434532/what-data-transfer-rates-are-needed-or-streaming-hd-1080p-or-720p-video-or-stan`, a 1 Gb/sec data rate is more than sufficient to meet the 22 Mbps transfer rate requirement for 1080p video used for many purposes today. This transfer rate might not be good enough for 4K video data rates as shown by the table found at `http://vashivisuals.com/4k-beyond-video-data-rates/`.

If you find that binary doesn't quite meet your video or audio processing needs, you can also try the cereal library (`http://hackage.haskell.org/package/cereal`). It provides many of the same features as binary, but uses a different coding strategy (strict versus lazy execution). You can read a short discussion of the differences at `https://stackoverflow.com/questions/14658031/cereal-versus-binary`.

GHC VERSION

Most of the libraries you use with Haskell will specify a GHC version. The version number tells you the requirements for the Haskell environment; the library won't work with an older GHC version. In most cases, you want to keep your copy of Haskell current to ensure that the libraries you want to use will work with it. Also, note that many library descriptions will include support requirements in addition to the version number. Often, you must perform GHC upgrades to obtain the required support or import other libraries. Make sure to always understand the GHC requirements before using a library or assuming that the library isn't working properly.

Hascore

The Hascore library found at `https://wiki.haskell.org/Haskore` gives you the means to describe music. You use this library to create, analyze, and manipulate music in various ways. An interesting aspect of this particular library is that it helps you see music in a new way. It also enables people who might not ordinarily be able to work with music express themselves. The site shows how the library makes lets you visualize music as a kind of math expression.

Of course, some musicians probably think that viewing music as a kind of math is to miss the point. However, you can find a wealth of sites that fully embrace the math in music, such as the American Mathematical Society (AMS) page at `http://www.ams.org/publicoutreach/math-and-music`. Some sites, such as Scientific American (`https://www.scientificamerican.com/article/is-there-a-link-between-music-and-math/`) even express the idea that knowing music can help someone understand math better, too.

REMEMBER

The point is that Hascore enables you to experience music in a new way through Haskell application programming. You can find other music and sound oriented libraries at `https://wiki.haskell.org/Applications_and_libraries/Music_and_sound`.

vect

Computer graphics in computers are based heavily in math. Haskell provides a wide variety of suitable math libraries for graphic manipulation, but vect (`http://hackage.haskell.org/package/vect`) represents one of the better choices because it's relatively fast and doesn't get mired in detail. Plus, you can find it used in existing applications such as the LambdaCube engine (`http://hackage.haskell.org/package/lambdacube-engine`), which helps you to render advanced graphics on newer hardware.

TIP

If your main interest in a graphics library is to experiment with relatively simple output, vect does come with OpenGL (`https://www.opengl.org/`) support, including projective four-dimensional operations and quaternions. You must load the support separately, but the support is fully integrated into the library.

vector

All sorts of programming tasks revolve around the use of arrays. The immutable built-in list type is a linked-list configuration, which means that it can use memory inefficiently and not process data requests at a speed that will work for your application. In addition, you can't pass a linked list to other languages, which may be a requirement when working in a graphics or other scenario in which high-speed interaction with other languages is a requirement. The vector library (http://hackage.haskell.org/package/vector) solves these and many other issues for which an array will work better than a linked list.

The vector library not only includes a wealth of features for managing data but also provides both mutable and immutable forms. Yes, using mutable data objects is the bane of functional programming, but sometimes you need to bend the rules a bit to process data fast enough to have it available when needed. Because of the nature of this particular library, you should see the need for eager execution (in place of the lazy execution that Haskell normally relies on) as essential. The use of eager processing also ensures that no potential for data loss exists and that cache issues are fewer.

aeson

A great many data stores today use JavaScript Object Notation (JSON) as a format. In fact, you can find JSON used in places you might not initially think about. For example, Amazon Web Services (AWS), among others, uses JSON to do everything from creating processing rules to creating configuration files. With this need in mind, you need a library to manage JSON data in Haskell, which is where aeson (http://hackage.haskell.org/package/aeson) comes into play. This library provides everything needed to create, modify, and parse JSON data in a Haskell application.

LIBRARY NAMES

Many of the library names in this chapter are relatively straightforward. For example, the text library works on text, so it's not hard to remember what to import when you use it. However, some library names are a bit more creative, which is the case with aeson. It turns out that in Greek mythology, Aeson is the father of Jason (http://www.argonauts-book.com/aeson.html). Of course, in this case, JSON did come first.

attoparsec

Mixed-format data files can present problems. For example, an HTML page can contain both ASCII and binary data. The attoparsec library (`http://hackage.haskell.org/package/attoparsec`) provides you with the means for parsing these complex data files and extracting the data you need from them. The actual performance of this particular library depends on how you write your parser and whether you use lazy evaluation. However, according to a number of sources, you should be able to achieve relatively high parsing speeds using this library.

TIP

One of the more interesting ways to use attoparsec is to parse log files. The article at `https://www.schoolofhaskell.com/school/starting-with-haskell/libraries-and-frameworks/text-manipulation/attoparsec` discusses how to use the library for this particular task. The article also gives an example of what writing a parser involves. Before you decide to use this particular library, you should spend time with a few tutorials of this type to ensure that you understand the parser creation process.

bytestring

You use the bytestring (`http://hackage.haskell.org/package/bytestring`) library to interact with binary data, such as network packets. One of the best things about using bytestring is that it allows you to interact with the data using the same features as Haskell lists. Consequently, the learning curve is less steep than you might imagine and your code is easier to explain to others. The library is also optimized for high performance use, so it should meet any speed requirements for your application.

Unlike many other parts of Haskell, bytestring also enables you to interact with data in the manner you actually need. With this in mind, you can use one of two forms of bytestring calls:

>> **Strict:** The library retains the data in one huge array, which may not use resources efficiently. However, this approach does let you to interact with other APIs and other languages. You can pass the binary data without concern that the data will appear fragmented to the recipient.

>> **Lazy:** The library uses smaller strict arrays to hold the data. This approach uses resources more efficiently and can speed data transfers. You use the lazy approach when performing tasks such as streaming data online.

TIP

The bytestring library also provides support for a number of data presentations to make it easier to interact with the data in a convenient manner. In addition, you can mix binary and character data as needed. A `Builder` module also lets you easily create byte strings using simple concatenation.

stringsearch

Manipulating strings can be difficult, but you're aided by the fact that the data you manipulate is in human-readable form for the most part. When it comes to byte strings, the patterns are significantly harder to see, and precision often becomes more critical because of the manner in which applications use byte strings. The stringsearch library (`http://hackage.haskell.org/package/stringsearch`) enables you to perform the following tasks on byte strings quite quickly:

>> Search for particular byte sequences

>> Break the strings into pieces using specific markers

>> Replace specific byte sequences with new sequences

This library will work with both strict and lazy byte strings. Consequently, it makes a good addition to libraries such as bytestring, which support both forms of bytestring calls. The page at `http://hackage.haskell.org/package/string search-0.3.6.6/docs/Data-ByteString-Search.html` tells you more about how this library performs its various tasks.

text

There are times when the text-processing capabilities of Haskell leave a lot to be desired. The text library (`http://hackage.haskell.org/package/text`) helps you to perform a wide range of tasks using text in various forms, including Unicode. You can encode or decode text as needed to meet the various Unicode Transformation Format (UTF) standards.

TIP

As helpful as it is to have a library for managing Unicode, the text library does a lot more with respect to text manipulation. For one thing, it can help you with internationalization issues, such as proper capitalization of words in strings.

This library also works with byte strings in both a strict and lazy manner (see the "bytestring" section, earlier in this chapter). Providing this functionality means that the text library can help you in streaming situations to perform text conversions quickly.

moo

The moo library (`http://hackage.haskell.org/package/moo`) provides Genetic Algorithm (GA) functionality for Haskell. GA is often used to perform various kinds of optimizations and to solve search problems using techniques found in nature (natural selection). Yes, GA also helps in understanding physical or natural environments or objects, as you can see in the tutorial at `https://towardsdata science.com/introduction-to-genetic-algorithms-including-example-code-e396e98d8bf3?gi=a42e35af5762`. The point is that it relies on evolutionary theory, one of the tenets of Artificial Intelligence (AI). This library supports a number of GA variants out of the box:

» Binary using bit-strings:

  • Binary and Gray encoding

  • Point mutation

  • One-point, two-point, and uniform crossover

» Continuous using a sequence of real values:

  • Gaussian mutation

  • BLX-α, UNDX, and SBX crossover

TIP

You can also create other variants through coding. These potential variants include

» Permutation

» Tree

» Hybrid encodings, which would require customizations

The readme (`http://hackage.haskell.org/package/moo-1.0#readme`) for this library tells you about other moo features and describes how they relate to the two out-of-the-box GA variants. Of course, the variants you code will have different features depending on your requirements. The single example provided with the readme shows how to minimize Beale's function (see `https://www.sfu.ca/~ssurjano/beale.html` for a description of this function). You may be surprised at how few lines of code this particular example requires.

Chapter **19**

Ten (Plus) Must-Have Python Packages

This chapter reviews just a few of the more interesting Python packages available today. Unlike with Haskell, finding reviews of Python packages is incredibly easy, along with articles stating people's lists of favorite packages. However, if you want to look at a more-or-less complete listing, the best place is the Python Package Index at https://pypi.org/. The list is so huge that you won't find a single list but must search through categories or for particular needs. Consequently, this chapter reflects just a few interesting choices, and if you don't see what you need, you really should search online.

MODULES, PACKAGES, AND LIBRARIES

There is general confusion over some terms (*module, package, and library*) used in Python and, unfortunately, this book won't help you untie this Gordian knot. When possible, this chapter uses the vendor term for whatever product you're reading about. However, the terms do have different meanings, which you can read about at `https://knowpapa.com/modpaclib-py/`. Consequently, sites such as PyPI use *package* (`https://pypi.org/`) because they offer collections of *modules* (which are individual `.py` files), while some vendors use the term *library*, presumably because the product uses compiled code created in another language, such as C.

Of course, you might ask why Python's core code is called the core library. That's because the core library is written in C and compiled, but then you have access to all the *packages* (collections of modules) that add to that core library. If you find that one or more of the descriptions in this chapter contain the wrong term, it's really not a matter of wanting to use the wrong term; it's more a of matter of dealing with the confusion caused by multiple terms that aren't necessarily well defined or appropriately used.

Gensim

Gensim (`https://radimrehurek.com/gensim/`) is a Python library that can perform natural language processing (NLP) and unsupervised learning on textual data. It offers a wide range of algorithms to choose from:

>> TF-IDF

>> Random projections

>> Latent Dirichlet allocation

>> Latent semantic analysis

>> Semantic algorithms:

- word2vec

- document2vec (`https://code.google.com/archive/p/word2vec/`)

TECHNICAL STUFF

Word2vec is based on neural networks (shallow, not deep learning, networks) and it allows meaningful transformations of words into vectors of coordinates that you can operate in a semantic way. For instance, operating on the vector representing `Paris`, subtracting the vector `France,` and then adding the vector `Italy` results in the vector `Rome`, demonstrating how you can use mathematics and the right Word2vec model to operate semantic operations on text. Fortunately, if this seems like Greek to you, Gensim offers excellent tutorials to make using this product easier.

PyAudio

One of the better platform-independent libraries to make sound work with your Python application is PyAudio (`http://people.csail.mit.edu/hubert/pyaudio/`). This library lets you record and play back sounds as needed. For example, a user can record an audio note of tasks to perform later and then play back the list of items as needed).

TIP

Working with sound on a computer always involves trade-offs. For example, a platform-independent library can't take advantage of special features that a particular platform might possess. In addition, it might not support all the file formats that a particular platform uses. The reason to use a platform-independent library is to ensure that your application provides basic sound support on all systems that it might interact with.

USING SOUND APPROPRIATELY

Sound is a useful way to convey certain types of information to the user. However, you must exercise care in using sound because special-needs users might not be able to hear it, and for those who can, using too much sound can interfere with normal business operations. However, sometimes audio is an important means of communicating supplementary information to users who can interact with it (or it can simply add a bit of pizzazz to make your application more interesting).

CLASSIFYING PYTHON SOUND TECHNOLOGIES

Realize that sound comes in many forms in computers. The basic multimedia services provided by Python (see the documentation at `https://docs.python.org/3/library/mm.html`) provide essential playback functionality. You can also write certain types of audio files, but the selection of file formats is limited. In addition, some packages, such as winsound (`https://docs.python.org/3/library/winsound.html`), are platform dependent, so you can't use them in an application designed to work everywhere. The standard Python offerings are designed to provide basic multimedia support for playing back system sounds.

The middle ground, augmented audio functionality designed to improve application usability, is covered by libraries such as PyAudio. You can see a list of these libraries at `https://wiki.python.org/moin/Audio`. However, these libraries usually focus on business needs, such as recording notes and playing them back later. Hi-fidelity output isn't part of the plan for these libraries.

Gamers need special audio support to ensure that they can hear special effects, such as a monster walking behind them. These needs are addressed by libraries such as PyGame (`http://www.pygame.org/news.html`). When using these libraries, you need higher-end equipment and have to plan to spend considerable time working on just the audio features of your application. You can see a list of these libraries at `https://wiki.python.org/moin/PythonGameLibraries`.

PyQtGraph

Humans are visually oriented. If you show someone a table of information and then show the same information as a graph, the graph is always the winner when it comes to conveying information. Graphs help people see trends and understand why the data has taken the course that it has. However, getting those pixels that represent the tabular information onscreen is difficult, which is why you need a library such as PyQtGraph (`http://www.pyqtgraph.org/`) to make things simpler.

Even though the library is designed around engineering, mathematical, and scientific requirements, you have no reason to avoid using it for other purposes. PyQtGraph supports both 2-D and 3-D displays, and you can use it to generate new graphics based on numeric input. The output is completely interactive, so a user can select image areas for enhancement or other sorts of manipulation. In addition, the library comes with a wealth of useful widgets (controls, such as buttons, that you can display onscreen) to make the coding process even easier.

REMEMBER

Unlike many of the offerings in this chapter, PyQtGraph isn't a free-standing library, which means that you must have other products installed to use it. This isn't unexpected because PyQtGraph is doing quite a lot of work. You need these items installed on your system to use it:

>> Python version 2.7 or higher

>> PyQt version 4.8 or higher (https://wiki.python.org/moin/PyQt) or PySide (https://wiki.python.org/moin/PySide)

>> numpy (http://www.numpy.org/)

>> scipy (http://www.scipy.org/)

>> PyOpenGL (http://pyopengl.sourceforge.net/)

TkInter

Users respond to the Graphical User Interface (GUI) because it's friendlier and requires less thought than using a command-line interface. Many products out there can give your Python application a GUI. However, the most commonly used product is TkInter (https://wiki.python.org/moin/TkInter). Developers like it so much because TkInter keeps things simple. It's actually an interface for the Tool Command Language (Tcl)/Toolkit (Tk) found at http://www.tcl.tk/. A number of languages use Tcl/Tk as the basis for creating a GUI.

TIP

You might not relish the idea of adding a GUI to your application. Doing so tends to be time consuming and doesn't make the application any more functional (it also slows down the application, in many cases). The point is that users like GUIs, and if you want your application to see strong use, you need to meet user requirements.

PrettyTable

Displaying tabular data in a manner the user can understand is important. Python stores this type of data in a form that works best for programming needs. However, users need something that is organized in a manner that humans understand and that is visually appealing. The PrettyTable library (https://pypi.python.org/pypi/PrettyTable) lets you easily add an appealing tabular presentation to your command-line application.

SQLAlchemy

A *database* is essentially an organized manner of storing repetitive or structured data on disk. For example, customer *records* (individual entries in the database) are repetitive because each customer has the same sort of information requirements, such as name, address, and telephone number. The precise organization of the data determines the sort of database you're using. Some database products specialize in text organization, others in tabular information, and still others in random bits of data (such as readings taken from a scientific instrument). Databases can use a tree-like structure or a flat-file configuration to store data. You'll hear all sorts of odd terms when you start looking into DataBase Management System (DBMS) technology — most of which will mean something only to a DataBase Administrator (DBA) and won't matter to you.

REMEMBER

The most common type of database is called a Relational DataBase Management System (RDBMS), which uses tables that are organized into records and fields (just like a table you might draw on a sheet of paper). Each *field* is part of a column of the same kind of information, such as the customer's name. Tables are related to each other in various ways, so creating complex relationships is possible. For example, each customer may have one or more entries in a purchase-order table, and the customer table and the purchase-order table are therefore related to each other.

An RDBMS relies on a special language called the Structured Query Language (SQL) to access the individual records inside. Of course, you need some means of interacting with both the RDBMS and SQL, which is where SQLAlchemy (`http://www.sqlalchemy.org/`) comes into play. This product reduces the amount of work needed to ask the database to perform tasks such as returning a specific customer record, creating a new customer record, updating an existing customer record, and deleting an old customer record.

Toolz

The Toolz package (`https://github.com/pytoolz/toolz`) fills in some of the functional programming paradigm gaps in Python. You specifically use it for functional support of

>> Iterators

>> Functions

>> Dictionaries

Interestingly enough, this same package works fine for both Python 2.*x* and 3.*x* developers, so you can get a single package to meet many of your functional data-processing needs. This package is a pure Python implementation, which means that it works everywhere.

TIP

If you need additional speed, don't really care about interoperability with every third-party package out there, and don't need the ability to work on every platform, you can use a Cython (`http://cython.org/`) implementation of the same package called CyToolz (`https://github.com/pytoolz/cytoolz/`). Besides being two to five times faster, CyToolz offers access to a C API, so there are some advantages to using it.

Cloudera Oryx

Cloudera Oryx (`http://www.cloudera.com/`) is a machine learning project for Apache Hadoop (`http://hadoop.apache.org/`) that provides you with a basis for performing machine learning tasks. It emphasizes the use of live data streaming. This product helps you add security, governance, and management functionality that's missing from Hadoop so that you can create enterprise-level applications with greater ease.

The functionality provided by Oryx builds on Apache Kafka (`http://kafka.apache.org/`) and Apache Spark (`http://spark.apache.org/`). Common tasks for this product are real-time spam filters and recommendation engines. You can download Oryx from `https://github.com/cloudera/oryx`.

funcy

The funcy package (`https://github.com/suor/funcy/`) is a mix of features inspired by clojure (`https://clojure.org/`). It allows you to make your Python environment better oriented toward the functional programming paradigm, while also adding support for data processing and additional algorithms. That sounds like a lot of ground to cover, and it is, but you can break the functionality of this particular package into these areas:

» Manipulation of collections

» Manipulation of sequences

» Additional support for functional programming constructs

- » Creation of decorators
- » Abstraction of flow control
- » Additional debugging support

TIP

Some people might skip the bottom part of the GitHub download pages (and for good reason; they normally don't contain a lot of information). However, pages the author of the funcy provides access to essays about why funcy implements certain features in a particular manner and those essay links appear at the bottom of the GitHub page. For example, you can read "Abstracting Control Flow" (http://hackflow.com/blog/2013/10/08/abstracting-control-flow/), which helps you understand the need for this feature, especially in a functional environment. In fact, you might find that other GitHub pages (not many, but a few) also contain these sorts of helpful links.

SciPy

The SciPy (http://www.scipy.org/) stack contains a host of other libraries that you can also download separately. These libraries provide support for mathematics, science, and engineering. When you obtain SciPy, you get a set of libraries designed to work together to create applications of various sorts. These libraries are:

- » NumPy
- » SciPy
- » matplotlib
- » IPython
- » Sympy
- » Pandas

The SciPy library itself focuses on numerical routines, such as routines for numerical integration and optimization. SciPy is a general-purpose library that provides functionality for multiple problem domains. It also provides support for domain-specific libraries, such as Scikit-learn, Scikit-image, and statsmodels. To make your SciPy experience even better, try the resources at http://www.scipy-lectures.org/. The site contains many lectures and tutorials on SciPy's functions.

XGBoost

The XGBoost package (https://github.com/dmlc/xgboost) enables you to apply a Gradient Boosting Machine (GBM) (https://towardsdatascience.com/boosting-algorithm-gbm-97737c63daa3?gi=df155908abce) to any problem, thanks to its wide choice of objective functions and evaluation metrics. It operates with a variety of languages, including

- » Python
- » R
- » Java
- » C++

TIP

In spite of the fact that GBM is a sequential algorithm (and thus slower than others that can take advantage of modern multicore computers), XGBoost leverages multithread processing in order to search in parallel for the best splits among the features. The use of multithreading helps XGBoost turn in an unbeatable performance when compared to other GBM implementations, both in R and Python. Because of all that it contains, the full package name is eXtreme Gradient Boosting (or XGBoost for short). You can find complete documentation for this package at https://xgboost.readthedocs.org/en/latest/.

Chapter **20**

Ten Occupation Areas that Use Functional Programming

F or many people, the reason to learn a new language or a new programming paradigm focuses on the ability to obtain gainful employment. Yes, they also have the joy of learning something new. However, to be practical, the something new must also provide a tangible result. The purpose of this chapter is to help you see the way to a new occupation that builds on the skills you discover through the functional programming paradigm.

Starting with Traditional Development

When asked about functional programming occupations, a number of developers who use functional programming in their jobs actually started with a traditional job and then applied functional programming methodologies to it. When coworkers saw that these developers were writing cleaner code that executed faster, they started adopting functional programming methodologies as well.

REMEMBER

Theoretically, this approach can apply to any language, but it helps to use a pure language (such as Haskell) when you can, or an impure language (such as Python) when you can't. Of course, you'll encounter naysayers who will tell you that functional programming applies only to advanced developers who are already working as programmers, but if that were the case, a person wouldn't have a place to start. Some organization will be willing to experiment with functional programming and continue to rely on it after the developers using it demonstrate positive results.

The problem is how to find such an organization. You can look online at places such as Indeed.com (`https://www.indeed.com/q-Haskell-Functional-Program ming-jobs.html`), which offers listings for the languages that work best for functional programming in traditional environments. At the time of this writing, Indeed.com had 175 Haskell job listings alone. Jobs for Python programmers with functional programming experience topped 6,020 (`https://www.indeed.com/ q-Python-Functional-Programming-jobs.html`).

TIP

A few websites deal specifically with functional programming jobs. For example, Functional Jobs (`https://functionaljobs.com/`) provides an interesting list of occupations that you might want to try. The benefit of these sites is that the listings are extremely targeted, so you know you'll actually perform functional programming. A disadvantage is that the sites tend to be less popular than mainstream sites, so you may not see the variety of jobs that you were expecting.

Going with New Development

With the rise of online shopping, informational, and other kinds of sites, you can bet that a lot of new development is also going on. In addition, traditional organizations will require support for new strategies, such as using Amazon Web Services (AWS) to reduce costs (see *AWS For Admins For Dummies* and *AWS For Developers For Dummies,* by John Paul Mueller [Wiley], for additional information on AWS). Any organization that wants to use serverless computing, such as AWS Lambda (`https://aws.amazon.com/lambda/`), will likely need developers who are conversant in functional programming strategies. Consequently, the investment in learning the functional programming paradigm can pay off in the form of finding an interesting job using new technologies rather than spending hour after boring hour updating ancient COBOL code on a mainframe.

TIP

When going the new development route, be sure you understand the requirements for your job and have any required certifications. For example, when working with AWS, your organization may require that you have an AWS Certified Developer (or other) certification. You can find the list of AWS certifications at https://aws.amazon.com/certification/. Of course, other cloud organizations exist, such as Microsoft Azure and Google Cloud. The article at https://www.zdnet.com/article/cloud-providers-ranking-2018-how-aws-microsoft-google-cloud-platform-ibm-cloud-oracle-alibaba-stack/ tells you about the relative strengths of each of these offerings.

Creating Your Own Development

Many developers started in their home or garage tinkering with things just to see what would happen. Becoming fascinated with code — its essence — is part of turning development into a passion rather than just a job. Some of the richest, best-known people in the world started out as developer entrepreneurs (think people like Jeff Bezos and Bill Gates). In fact, you can find articles online, such as the one at https://skillcrush.com/2014/07/15/developers-great-entrepreneurs/, that tell precisely why developers make such great entrepreneurs. The advantage of being your own boss is that you do things your way, make your mark on the world, and create a new vision of what software can do.

Yes, sometimes you get the money, too, but more developers have found that they become successful only after they figure out that creating your own development environment is all about business — that is, offering a service that someone else will buy. Articles, such as the one at https://hackernoon.com/reality-smacking-tips-to-help-you-transition-from-web-developer-to-entrepreneur-9644a5cbe0ff and https://codeburst.io/the-walk-of-becoming-a-software-developer-entrepreneur-ef16b50bab76, tell you how to make the transition from developer to entrepreneur.

REMEMBER

The functional connection comes into play when you start to consider that the functional programming paradigm is somewhat new. Businesses are starting to pay attention to functional programming because of articles such as the Info-World offering at https://www.infoworld.com/article/3190185/software/is-functional-programming-better-for-your-startup.html. When businesses find out that functional programming not only creates better code but also makes developers more productive (see the article at https://medium.com/@xiaoyunyang/why-functional-programming-from-a-developer-productivity-perspective-69c4b8100776), they begin to see a financial reason to employ consultants (that's you) to move their organizations toward the functional programming paradigm.

Finding a Forward-Thinking Business

Many businesses are already using functional programming methodologies. In some cases, these businesses started with functional programming, but in more cases the business transitioned. One such business is Jet.com (https://jet.com/), which offers online shopping that's like a mix of Amazon.com (https://www.amazon.com/) and Costco (https://www.costco.com/). You can read about this particular business at https://www.kiplinger.com/article/spending/T050-C011-S001-what-you-need-to-know-before-joining-jet-com.html. The thing that will interest you is that Jet.com relies on F#, a multiparadigm language similar to Python from an environmental perspective, to meet its needs.

TIP

Most languages want you to know that real companies are using them to do something useful. Consequently, you can find a site that provides a list of these organizations, such as https://wiki.haskell.org/Haskell_in_industry for Haskell and https://wiki.python.org/moin/OrganizationsUsingPython for Python. Languages that are more popular will also sprout a lot of articles. For example, the article at https://realpython.com/world-class-companies-using-python/ supplies a list of well-known organizations that use Python. You need to exercise care in applying to these organizations, however, because you never know whether you'll actually work with your programming language of choice (or whether you'll work as a developer at all).

Doing Something Really Interesting

Some people want to go to work, do a job for eight to ten hours, and then come home and forget about work. This section isn't for you. On the flip side, some people want to make their mark on the world and light it on fire. This section won't work for you, either. This section is for those people who fall between these two extremes: Those who don't mind working a few extra hours as long as the work is interesting and meaningful, and they don't have to manage any business details. After all, the fun of functional programming is writing the code and figuring out interesting ways to make data jump through all sorts of hoops. That's where job sites like Functional Works (https://functional.works-hub.com/) come into play.

REMEMBER

Sites such as Functional Works search for potential candidates for large organizations, such as Google, Facebook, Two Sigma, and Spotify. The jobs are listed by category in most cases. Be prepared to read for a while because the sites generally describe the jobs in detail. That's because these organizations want to be sure that you know what you're getting into, and they want to find the best possible fit.

These sites often offer articles, such as "Compose Tetras" (`https://functional.works-hub.com/learn/compose-tetris-61b59`). The articles are interesting because they give you a better perspective of what the site is about, and why a company would choose this site, rather than another one, to find people. You learn more about functional programming, as well.

Developing Deep Learning Applications

One of the most interesting and widely discussed subsets of Artificial Intelligence (AI) today is that of deep learning, in which algorithms use huge amounts of data to discover patterns and then use those patterns to perform data-based tasks. You might see the output as being voice recognition or robotics, but the computer sees data — lots and lots of data. Oddly enough, functional programming techniques make creating deep learning applications significantly easier, as described in the article at `https://towardsdatascience.com/functional-programming-for-deep-learning-bc7b80e347e9?gi=1f073309a77c`. This article is interesting because it looks at a number of languages that aren't discussed in this book but are just as important in the world of functional programming. You can learn more about the world of AI in *AI For Dummies*, by John Paul Mueller and Luca Massaron (Wiley), and the world of machine learning in *Machine Learning For Dummies*, also by John Paul Mueller and Luca Massaron (Wiley).

Writing Low-Level Code

You might not initially think about using functional programming methods to write low-level code, but the orderly nature of functional programming languages makes them perfect for this task. Here are a few examples:

>> **Compilers and interpreters:** These applications (and that's what they are) work through many stages of processing, relying on tree-like structures to turn application code into a running application. Recursion makes processing tree-like structures easy, and functional languages excel at recursion (see the article at `https://stackoverflow.com/questions/2906064/why-is-writing-a-compiler-in-a-functional-language-easier` for details). The Compcert C Compiler (`http://compcert.inria.fr/compcert-C.html`) is one example of this use.

>> **Concurrent and parallel programming:** Creating an environment in which application code executes concurrently, in parallel, is an incredibly hard task for most programming languages, but functional languages handle this task

with ease. You could easily write a host environment using a functional language for applications written in other languages.

>> **Security:** The immutable nature of functional code makes it inherently safe. Creating the security features of an operating system or application using functional code significantly reduces the chance that the system will be hacked.

You can more easily address a wide range of low-level coding applications in a functional language because of how functional languages work. A problem can arise, however, when resources are tight because functional languages can require more resources than other languages. In addition, if you need real-time performance, a functional language may not provide the ultimate in speed.

Helping Others in the Health Care Arena

The health care field is leading the charge in creating new jobs, so your new job might just find you in the health care industry, according to the article at `https://www.cio.com/article/2369526/careers-staffing/103069-10-Hottest-Healthcare-IT-Developer-and-Programming-Skills.html`. If you regard working in the medical industry as possibly the most boring job in the world, read ads like the one at `https://remoteok.io/remote-jobs/64883-remote-functional-programming-medical-systems`. The possibilities might be more interesting than you think. Oddly enough, many of these ads, the one referenced in this paragraph included, specifically require you to have functional programming experience. This particular job also specifies that the job environment is relaxed and the company expects you to be innovative in your approach to solving problems — which is hardly a formula for a boring job.

Working as a Data Scientist

As a data scientist, you're more likely to use the functional programming features of Python than to adapt a wholly functional approach by using a language such as Haskell. According to the article at `https://analyticsindiamag.com/top-10-programming-languages-data-scientists-learn-2018/`, Python is still the top language for data science.

Articles such as the one at `https://www.kdnuggets.com/2015/04/functional-programming-big-data-machine-learning.html` seem to question just how much penetration functional programming has made in the data science community; however, such penetration exists. The discussion at `https://datascience.stackexchange.com/questions/30578/what-can-functional-programming-be-used-for-in-data-science` details good reasons for data scientists to use functional programming, including better ways to implement parallel programming. When you consider that a data scientist could rely on a GPU with up to 5,120 cores (such as the NVidia Titan V, `https://www.nvidia.com/en-us/titan/titan-v/`), parallel programming takes on a whole new meaning.

REMEMBER

Of course, data science involves more than just analyzing huge datasets. The act of cleaning the data and making the various data sources work together is extremely time consuming, especially in getting the various data types aligned. However, even in this regard, using a functional language can be an immense help. Knowing a functional language gives you an edge as a data scientist — one that could lead to advancement or more interesting projects that others without your edge will miss. The book *Python For Data Science For Dummies,* by John Paul Mueller and Luca Massaron (Wiley), provides significant insights into just how you can use Python to your advantage in data science, and implementing functional programming techniques in Python is just another step beyond.

Researching the Next Big Thing

Often you'll find a query for someone interested in working as a researcher on a job site such as Indeed.com (`https://www.indeed.com/`). In some cases, the listing will specifically state that you need functional programming skills. This requirement exists because working with huge datasets to determine whether a particular process is possible or an experiment succeeded, or to get the results of the latest study, all demand strict data processing. By employing functional languages, you can to perform these tasks quickly using parallel processing. The strict typing and immutable nature of functional languages are a plus as well.

TIP

Oddly enough, the favored languages for research, such as Clojure (see `https://www.theinquirer.net/inquirer/feature/2462362/7-new-generation-programming-languages-you-should-get-to-know`), are also the highest-paying languages, according to sites such as TechRepublic (`https://www.techrepublic.com/article/what-are-the-highest-paid-jobs-in-programming-the-top-earning-languages-in-2017/`). Consequently, if you want an interesting job in an incredibly competitive field with high pay, being a researcher with functional programming skills may be just what you're looking for.

Index

Symbols

- (dash), use in switches, 198
-- (double dash), use in switches, 198
% (percent sign), 192
%% (double percent sign), 192
%%timeit magic function, 194
%alias magic function, 193
%autocall magic function, 193
%automagic magic function, 193
%autosave magic function, 193
%cd magic function, 193
%cls magic function, 193
%colors magic function, 193
%config magic function, 193
%dhist magic function, 193
%file magic function, 194
%hist magic function, 194
%install_ext magic function, 194
%load magic function, 194
%load_ext magic function, 194
%lsmagic magic function, 194
%magic magic function, 194
%matplotlib magic function, 194, 238
%paste magic function, 194
%pdef magic function, 194
%pdoc magic function, 194
%pinfo magic function, 194
%pinfo2 magic function, 194
%reload_ext magic function, 194
%source magic function, 194
%timeit magic function, 194
(>>) monad sequencing operator, 195
. (dot) operator, 152, 158–159

/ (slash), use in switches, 198
:set command, 201
:t expression, 162, 167
:t openFile function, 190
:unset command, 200
\x0A output, 228
\x0D output, 228
__getattribute() function, 267
` (back quotation mark), use in Haskell, 98
| (OR) operator, 166–167
++ operator, 189
<<loop>> output, 256
<> operator, 173
== (equality) operator, 179

A

α (alpha)-conversion, 85
abstracting
  patterns, 112
  simply-typed calculus, 83–84
  untyped lambda calculus, 82–83
aeson library, 274
AI (Artificial Intelligence), 277
algorithms
  human element in, 251
  pattern matching, 118
  types of, 280
alpha (α)-conversion, 85
Amazon Web Services. *See* AWS
Amazon.com, 292
American Mathematical Society (AMS), 273
Anaconda
  add-ons for, 22
  applications within, 26–27

downloading, 21–22
installing
  on Linux, 22–23
  on Mac OS, 23–24
  on Windows, 24–26
version 5.1, 19
anchors, 115–116
AND operator, creating types with, 164–166
Apache Hadoop, 285
Apache Kafka, 285
Apache Spark, 285
append function, 75, 102
appendFile function, 191
appending data, 75, 102, 129–130, 173, 212
AppendMode argument, 212
applications
  Anaconda, 26–27
  Haskell, 56–59
  in lambda calculus, 81–82
  patterns in, 112
  Python, 34–38
  repetition in, 126–127
  state of, 13, 74–76
apply operators
  interacting with user, 189
  lambda calculus, 81–82
  mapping tasks, 152
Argparse library, 205, 206
args property, 266
arguments
  command line
    Haskell, 200–201
    Python, 205–206
  exception, 266–267
  replacing bound variables with, 86–87

lookup function, 133

loops
  defined, 250
  infinite, 250
  using recursive functions
    instead of, 127–128

low-level code, writing
  compilers, 293
  concurrent programming, 293
  interpreters, 293
  parallel programming, 293

M

machine code, 13

MacOS
  installing Anaconda on, 23–24
  installing Haskell on, 50–51

magic functions, Jupyter
  Notebook, 192–194

main function, 216

make_ blobs() function, 238

manipulating data
  I/O data, 144–145
  lists, 100–102
  types of data manipulation,
    144–145

map function, 72, 152

map object, 153

Map operator, 173

marks, types of, 98

math. See also lambda calculus
  abstraction, 171
  Peano arithmetic, 79
  solving problems with, 10

MATLAB, 10, 21

Maybe value, 174–176

measures, statistical, 180–181

membership, list, 101

MemoryError exception, 265

Microsoft Azure, 290

Miranda programming
  language, 16

mod operator, 156

modularization, 11

modules, 55–56, 280

monads
  functions, 195–196
  handles and, 188
  monad sequencing, 195

monoids, 170–174

monolithic code, 187

moo library, 277

multivariate regression, 237

music oriented libraries, 273

mutable variables, avoiding, 250

mypy static type checker, 163

N

n_samples argument, 238

--name argument, 204

name=Sam argument, 205

Native.py argument, 205

natural language processing
  (NLP), 280

Navigator, Anaconda, 26, 45–46

neoVim text editor, 49

NLP (natural language
  processing), 280

non-curried functions, 72, 74–76

non-strict (lazy) bytestring
  call, 275

non-strict (lazy) evaluation, 16

non-strict (lazy) language,
  251–252

notebooks. See also Jupyter
  Notebook
  creating new, 30–31
  deleting, 32
  exporting, 31–32
  importing, 32–33
  text editors versus, 21

Nothing value, 174–176

null values, 178–179

numeric values, 220

NumPy
  arrays, 221, 236
  evaluating lists with, 99
  slicing and dicing data
    with, 146

numpy/routines.io
  library, 236

O

Object-Oriented Programming
  (OOP), 14, 180

OCaml programming
  language, 12

occupations, using functional
  programming
  businesses, finding, 292
  data scientist, 295
  deep learning applications, 293
  developers, 289–291
  health care industry, 294–295
  low-level code writing, 293–294
  overview, 289
  researching, 295
  sites for, 292–293

OOP (Object-Oriented
  Programming), 14, 180

openBinaryFile function, 226

operations. See also lambda
  calculus
  of lambda calculus, 80–84
  reduction operations. See
    reduction operations

operator module, 160

OR (|) operator, 166–167

OS exceptions, 265

output. See also I/O
  comments as, 42
  inputs and, 17
  Int values as, 135
  in Jupyter Notebook, 30–31
  lines of text, multiple, 39–40
  output devices, 190
  procedures and, 67

P

packages
  defined, 280
  Haskell, 201–202
  Python, 279–287
pandas.io library, 235
paradigms, 9. *See also* functional programming paradigm
parallel programming, 293
parameterizing, 176–178
parentheses, in lambda calculus, 81
Paris vector, 281
partial file lock, 211
passing by reference, Python variables, 74–76
pattern matching
  algorithm, 118
  in analysis, 117–118
  in data, patterns, 112–113
  defined, 111
  in Haskell, 118–120
  in Python, 121–124
  regular expressions, 113–124
Peano arithmetic, 79
percent sign (%) symbol, 192
Posix matches, 118–119
Post, Emil Leon, 78
Prelude.catch function, 258
PrettyTable library, 283
prime (`) symbol, use in Haskell, 98
print() call, 237
print() function, 13–14, 30, 34, 210
programming. *See also* functional programming paradigm
  concurrent, 293
  declarative, 14
  functional, 9–18
  imperative, 13
  literate, 21
  object-oriented, 14

parallel, 293
procedural, 13–14
programming languages
  C#, 12
  C++, 12, 14, 162, 170
  functional features in, 12
  η-conversion and, 88
  impure, 17
  Lisp, 12
  Miranda, 16
  non functional, using type signatures, 162
  OCaml, 12
  pure, 16–17
  Racket, 12
  supporting coding styles, 17
  supporting coding styles with, 17
Prompt, Anaconda
  code, running using, 38
  description of, 27
  Infix module, installing with, 90
  lists, creating using, 95
prompts, in applications, 198–199
protocol, communication, 192
pure functions, 186, 187
pure languages, 11–12, 16–17
putCharUtf8 function, 226
putStringUtf8 function, 226
putStrLn function, 188–189, 190, 195, 227
PyAudio library, 281–282
PyGame library, 282
PyMonad library, 176
PyQtGraph library, 282–283
Python
  accessing command line in, 205–206
  Anaconda, 21–27
  bit manipulation and, 224
  calls, 215
  code, 28–33, 38–39
  comments in, 41–44

creating application, 34–38
data
  dicing with, 150–151
  filtering, 156
  immutable, 68
  mapping, 153–154
  organizing, 159–160
data typing, 137
datasets in, 33–34
errors in, 259–268
exception arguments, 266–267
functions in, 73–76, 139–140
as impure language, 17
indentation in, 39–41
interacting with binary data in, 228–229
Jupyter Notebook, 20–21, 44
lambda functions, creating in, 89–90
lists in, 95–96, 99–100, 102, 136–137
matches for, 121–123
objects, mutability of, 76
online resources for, 45–46
packages, 279–287
parameterizing types in, 178
pattern matching in, 121–124
reading data, 215
slicing data, 150–151
sound technologies of, 282
string-related tasks in, 106–107
type signatures, 163
Python Package Index, 279

Q

Quora website, 18

R

Racket programming language, 12
range, lambda calculus, 83
RCV1 (Reuters Corpus Volume I), 240

About the Author

John Mueller is a freelance author and technical editor. He has writing in his blood, having produced 110 books and more than 600 articles to date. The topics range from networking to artificial intelligence and from database management to heads-down programming. Some of his current books include discussions of data science, machine learning, and algorithms, all of which use Python as a demonstration language. His technical editing skills have helped more than 70 authors refine the content of their manuscripts. John has provided technical editing services to various magazines, performed various kinds of consulting, and writes certification exams. Be sure to read John's blog at http://blog.johnmuellerbooks.com/. You can reach John on the Internet at John@JohnMuellerBooks.com. John is also has a website at http://www.johnmuellerbooks.com/.

Dedication

This book is in remembrance of my niece Heather.

Acknowledgments

Thanks to my wife, Rebecca. Even though she is gone now, her spirit is in every book I write, in every word that appears on the page. She believed in me when no one else would.

Russ Mullen deserves thanks for his technical edit of this book—especially in dealing with what turned out to be a completely different sort of computer science topic. He greatly added to the accuracy and depth of the material you see here. Russ is always providing me with great URLs for new products and ideas. However, it's the testing that Russ does that helps most. He's the sanity check for my work. Russ also has different computer equipment from mine, so he's able to point out flaws that I might not otherwise notice.

Matt Wagner, my agent, deserves credit for helping me get the contract in the first place and taking care of all the details that most authors don't really consider. I always appreciate his assistance. It's good to know that someone wants to help.

A number of people read all or part of this book to help me refine the approach, test the coding examples, and generally provide input that all readers wish they could have. These unpaid volunteers helped in ways too numerous to mention here. I especially appreciate the efforts of Eva Beattie, Glenn A. Russell, Luca Massaron, and Osvaldo Téllez Almirall, who provided general input, read the entire book, and selflessly devoted themselves to this project.

Finally, I would like to thank Katie Mohr, Susan Christophersen, and the rest of the editorial and production staff.

Publisher's Acknowledgments

Associate Publisher: Katie Mohr

Project and Copy Editor: Susan Christophersen

Technical Editor: Russ Mullen

Sr. Editorial Assistant: Cherie Case

Production Editor: Mohammed Zafar Ali

Cover Image: © Henrik5000/iStock.com